endorsed for edexcel

Edexcel GCSE (9-1) Mathematics Higher EXTENSION

Practice, Reasoning and Problem-solving Book

Confidence • Fluency • Problem-solving • Reasoning

Author: Glyn Payne

PEARSON

Published by Pearson Education Limited, 80 Strand, London WC2R 0RL.

www.pearsonschoolsandfecolleges.co.uk

Copies of official specifications for all Edexcel qualifications may be found on the website: www.edexcel.com

Edited by ProjectOne Publishing Solutions, Scotland
Typeset and illustrated by Tech-set Ltd, Gateshead

First published 2015

18 17 16
10 9 8 7 6 5 4 3 2

British Library Cataloguing in Publication Data
A catalogue record for this book is available from the British Library

ISBN 978 1 292 10505 5

Printed in Slovakia by Neografia

Acknowledgements
The publisher would like to thank the following for their kind permission to reproduce their photographs:

Cover images: Front: Created by **Fusako**, Photography by NanaAkua

Every effort has been made to contact copyright holders of material reproduced in this book. Any omissions will be rectified in subsequent printings if notice is given to the publishers.

A note from the publisher

In order to ensure that this resource offers high-quality support for the associated Pearson qualification, it has been through a review process by the awarding body. This process confirms that this resource fully covers the teaching and learning content of the specification or part of a specification at which it is aimed. It also confirms that it demonstrates an appropriate balance between the development of subject skills, knowledge and understanding, in addition to preparation for assessment.

Endorsement does not cover any guidance on assessment activities or processes (e.g. practice questions or advice on how to answer assessment questions), included in the resource nor does it prescribe any particular approach to the teaching or delivery of a related course.

While the publishers have made every attempt to ensure that advice on the qualification and its assessment is accurate, the official specification and associated assessment guidance materials are the only authoritative source of information and should always be referred to for definitive guidance.

Pearson examiners have not contributed to any sections in this resource relevant to examination papers for which they have responsibility.

Examiners will not use endorsed resources as a source of material for any assessment set by Pearson.

Endorsement of a resource does not mean that the resource is required to achieve this Pearson qualification, nor does it mean that it is the only suitable material available to support the qualification, and any resource lists produced by the awarding body shall include this and other appropriate resources.

Contents

Welcome to Edexcel GCSE (9-1) Mathematics Higher Extension Practice, Reasoning and Problem-solving Book

This Extension Practice Book is packed with extra practice on all the new and most demanding content of the new GCSE Specification for Higher tier giving you more opportunities to practise answering questions to gain confidence and develop problem-solving and reasoning skills.

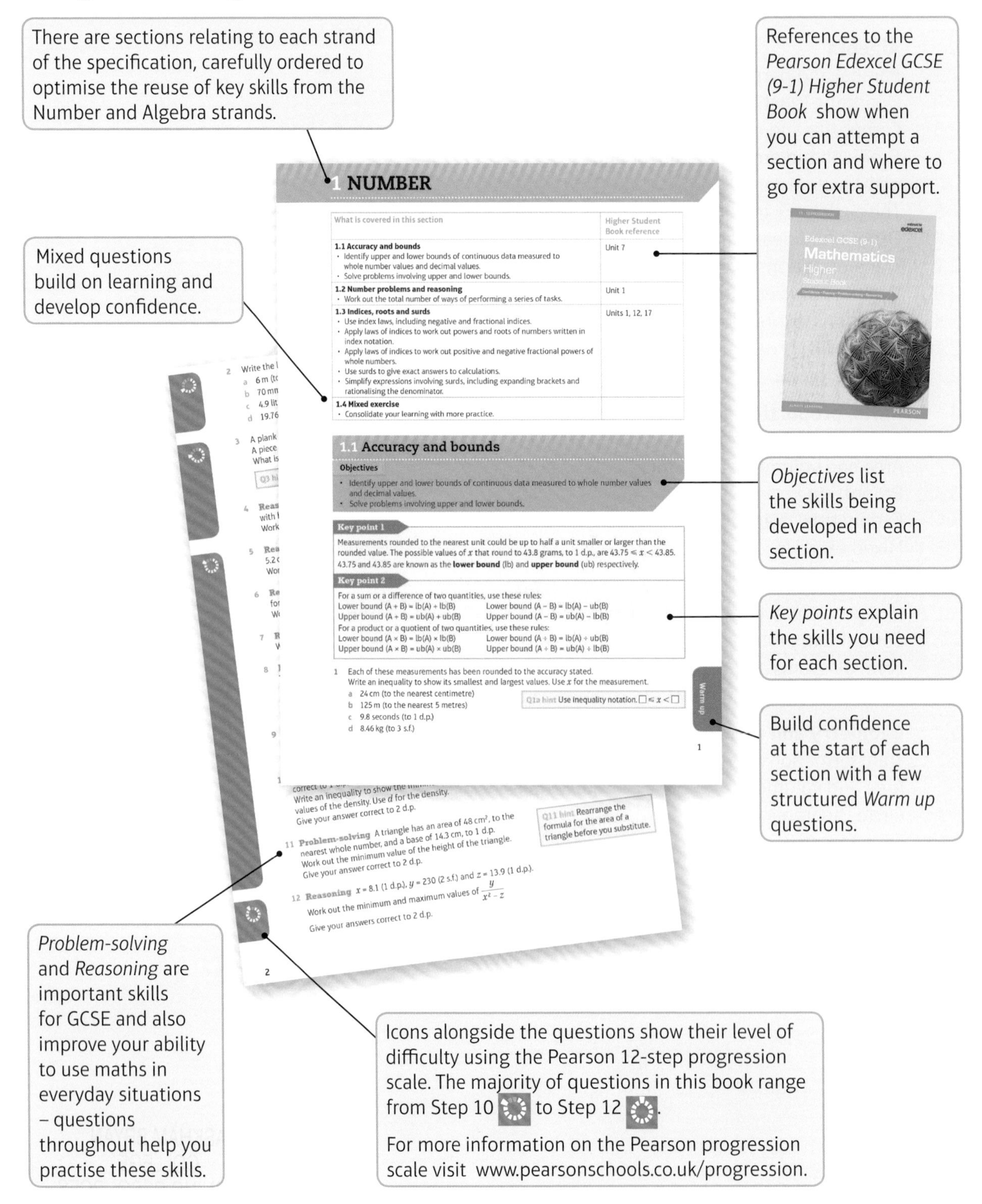

1 NUMBER

What is covered in this section	Higher Student Book reference
1.1 Accuracy and bounds • Identify upper and lower bounds of continuous data measured to whole number values and decimal values. • Solve problems involving upper and lower bounds.	Unit 7
1.2 Number problems and reasoning • Work out the total number of ways of performing a series of tasks.	Unit 1
1.3 Indices, roots and surds • Use index laws, including negative and fractional indices. • Apply laws of indices to work out powers and roots of numbers written in index notation. • Apply laws of indices to work out positive and negative fractional powers of whole numbers. • Use surds to give exact answers to calculations. • Simplify expressions involving surds, including expanding brackets and rationalising the denominator.	Units 1, 12, 17
1.4 Mixed exercise • Consolidate your learning with more practice.	

1.1 Accuracy and bounds

Objectives

- Identify upper and lower bounds of continuous data measured to whole number values and decimal values.
- Solve problems involving upper and lower bounds.

Key point 1

Measurements rounded to the nearest unit could be up to half a unit smaller or larger than the rounded value. The possible values of x that round to 43.8 grams, to 1 d.p., are $43.75 \leqslant x < 43.85$.
43.75 and 43.85 are known as the **lower bound** (lb) and **upper bound** (ub) respectively.

Key point 2

For a sum or a difference of two quantities, use these rules:

Lower bound (A + B) = lb(A) + lb(B) Lower bound (A – B) = lb(A) – ub(B)
Upper bound (A + B) = ub(A) + ub(B) Upper bound (A – B) = ub(A) – lb(B)

For a product or a quotient of two quantities, use these rules:

Lower bound (A × B) = lb(A) × lb(B) Lower bound (A ÷ B) = lb(A) ÷ ub(B)
Upper bound (A × B) = ub(A) × ub(B) Upper bound (A ÷ B) = ub(A) ÷ lb(B)

1 Each of these measurements has been rounded to the accuracy stated.
Write an inequality to show its smallest and largest values. Use x for the measurement.

a 24 cm (to the nearest centimetre)
b 125 m (to the nearest 5 metres)
c 9.8 seconds (to 1 d.p.)
d 8.46 kg (to 3 s.f.)

Q1a hint Use inequality notation. $\square \leqslant x < \square$

Warm up

2 Write the lower and upper bounds for each of these measurements.

a 6 m (to the nearest metre)

b 70 mm (to the nearest 10 mm)

c 4.9 litres (to 1 d.p.)

d 19.76 seconds (to 2 d.p.)

3 A plank of wood measures 2 metres, to the nearest cm.
A piece of length 83.5 cm, correct to 1 d.p. is sawn off.
What is the minimum length of the piece of wood that is left?

Q3 hint Minimum length remaining = minimum original length – maximum length sawn off

4 **Reasoning** A football pitch measures 106 m by 78 m, with both lengths measured to the nearest metre.
Work out the maximum perimeter of the pitch.

Q4 A common error is to use the wrong level of accuracy in the working and in the answer to the question.

5 **Reasoning** A rectangle has a length of 8.5 cm and a width of 5.2 cm, both given to the nearest millimetre.
Work out the lower and upper bounds for the area of the rectangle.

Q5 hint Use the rules for working out the lower and upper bounds of a product.

6 **Reasoning** A vehicle is travelling at a constant speed of 23.6 m/s, correct to 1 d.p., for 18 seconds, correct to the nearest second.
Work out the upper bound for the distance it will travel.

7 **Reasoning** A cube has a side length of 7.2 cm, correct to 1 d.p.
What is the smallest possible volume of the cube?

8 **Problem-solving** There are 12 identical marbles in a bag.
The bag weighs 16 g, correct to the nearest gram.
Each marble weighs 67 g, correct to the nearest gram.
What is the difference between the minimum and maximum possible weights of the bag of marbles?

Q8 A common error is to find all values and then not state which are the upper and lower bounds of the calculation.

9 **Reasoning** $x = 23.6$, $y = 9.4$ and $z = 18.7$, all correct to 1 d.p.

a Work out the minimum value of $2x - 3y$

b Work out the maximum value of $z^2 - xy$

10 **Reasoning** A gold bar has a volume of 12.4 cm^3, correct to 1 d.p. and a mass of 239 g, correct to 3 s.f.
Write an inequality to show the minimum and maximum values of the density. Use d for the density.
Give your answer correct to 2 d.p.

Q10 hint This is a division. Check the rules for maximum and minimum values.

11 **Problem-solving** A triangle has an area of 48 cm^2, to the nearest whole number, and a base of 14.3 cm, to 1 d.p.
Work out the minimum value of the height of the triangle.
Give your answer correct to 2 d.p.

Q11 hint Rearrange the formula for the area of a triangle before you substitute.

12 **Reasoning** $x = 8.1$ (1 d.p.), $y = 230$ (2 s.f.) and $z = 13.9$ (1 d.p.).

Work out the minimum and maximum values of $\frac{y}{x^2 - z}$

Give your answers correct to 2 d.p.

1.2 Number problems and reasoning

Objective

- Work out the total number of ways of performing a series of tasks.

Key point 3

When choice A can be made in m ways and choice B can be made in n ways, the total number of ways of choosing A then B is $m \times n$

A **factorial** is the result of multiplying a sequence of descending integers.

The $x!$ button on your calculator is the 'factorial' button. For example, $4! = 4 \times 3 \times 2 \times 1 = 24$

Warm up

1 **Problem-solving** A key pad has the digits 0 to 9 on it.

a How many ways can you select a 4-digit code if repetitions are allowed?

b How many ways can you select a 4-digit code if repetitions are not allowed?

2 **Problem-solving** Tom wants to make a code for his tablet. He decides to select two letters and three digits. He has a choice of the letters A, B, C and D, together with the digits 1 to 5. Repeat letters and/or digits are not allowed.

a How many possible codes are there?

b He decides to select two odd digits and one even digit. How many codes are now possible?

c How many codes would be possible if repetitions were allowed?

Q2 hint Work out how many ways the letters can be chosen, then how many ways the digits can be chosen, using the $m \times n$ rule.

3 **Problem-solving** Car registration plates used to consist of three letters, followed by three digits, followed by one letter.
How many different registration plates were possible if repetitions were not allowed in the first three letters or in the three digits?

4 **Problem-solving** You can you use the digits 1, 2, 3, 5, 7 and 8.

a How many 3-digit numbers can you make if repetitions are not allowed?

b How many of these 3-digit numbers will be odd?

c How many of these 3-digit numbers will be less than 200?

Q4b hint Select the last digit first because you need the number to be odd. How many ways can you do this? Then select the other digits.

Q4c hint For the number to be less than 200, what digit must it start with? Then select the other digits.

5 **Problem-solving** A restaurant offers 4 starters, 6 main courses and 5 desserts.

a If customers can choose from all three courses, how many ways can this be done?

b If customers can choose either a starter and a main course or a main course and a dessert, how many different options are possible?

6 **Reasoning** There are six points, A, B, C, D, E and F, on the circumference of a circle.
How many chords (for example, AD) can be drawn?

Q6 hint Remember that AB is the same chord as BA.

7 **Problem-solving** The letters of the word TRIANGLE are rearranged at random.
In how many of the possible arrangements will the first and last letters be vowels?

8 **Problem-solving** Tess has to make a 3-digit number using the digits 1, 2, 3, 4, 5, 6 and 7.
Repetitions are not allowed.

a How many ways can she make a number between 200 and 500?

b How many ways can she make an even number greater than 600?

1.3 Indices, roots and surds

Objectives

- Use index laws, including negative and fractional indices.
- Apply laws of indices to work out powers and roots of numbers written in index notation.
- Apply laws of indices to work out positive and negative fractional powers of whole numbers.
- Use surds to give exact answers to calculations.
- Simplify expressions involving surds, including expanding brackets and rationalising the denominator.

Key point 4

Use these **rules of indices** to multiply, divide and work out a power of a power:

$x^m \times x^n = x^{m+n}$ $\frac{x^m}{x^n} = x^{m-n}$ $(x^m)^n = x^{mn}$

Use these rules of indices to work out zero, negative and fractional powers:

$x^0 = 1$ $x^{-n} = \frac{1}{x^n}$ $x^{\frac{1}{n}} = \sqrt[n]{x}$ $x^{\frac{n}{m}} = (\sqrt[m]{x})^n$

Key point 5

Use these rules to multiply and divide **surds**:

$\sqrt{mn} = \sqrt{m} \times \sqrt{n}$ $\sqrt{\frac{m}{n}} = \frac{\sqrt{m}}{\sqrt{n}}$

Use these methods to **rationalise a denominator** to give the denominator as an integer:

$$\frac{a}{\sqrt{b}} = \frac{a}{\sqrt{b}} \times \frac{\sqrt{b}}{\sqrt{b}} = \frac{a\sqrt{b}}{b}$$

$$\frac{1}{a\sqrt{b}} = \frac{1}{a\sqrt{b}} \times \frac{\sqrt{b}}{\sqrt{b}} = \frac{\sqrt{b}}{ab}$$

$$\frac{c}{d+\sqrt{e}} = \frac{c}{d+\sqrt{e}} \times \frac{d-\sqrt{e}}{d-\sqrt{e}} = \frac{c(d-\sqrt{e})}{d^2-e}$$

Warm up

1 Work out the value of

a $2^{-5} \times 2^{12}$ b $(3^{-7})^0$ c $3^4 \div 3^{-1}$
d $(5^{-3})^{-1}$ e $(2^{-4})^2$ f $\sqrt{5^8}$
g $\sqrt[3]{10^9}$ h $\sqrt[8]{4^{32}}$

Q1f hint Rewrite the roots as a fractional index before you simplify. $\sqrt{5^8} = (5^8)^{\square} = \square$

2 Expand and simplify

a $(x-3)(x+8)$ b $(3x-5)(x-2)$
c $(2x+7)^2$ d $(x-2y)(4x+y)$

Q2a hint $(x-3)(x+8)$

3 Work out the value of

a $16^{\frac{1}{4}}$ b $(-216)^{\frac{1}{3}}$ c $64^{\frac{3}{2}}$ d $625^{\frac{3}{4}}$

4 Work out the value of

a $81^{-\frac{1}{2}}$ b $25^{-1.5}$ c $343^{-\frac{1}{3}}$
d $(-64)^{\frac{1}{3}} \times 3^{-2}$ e $100^{1.5} \times 5^{-3}$

5 Work out the value of

a $\left(\frac{64}{343}\right)^{-\frac{2}{3}}$ b $\left(1\frac{7}{9}\right)^{-\frac{3}{2}}$

c $\left(2\frac{1}{4}\right)^{-2.5}$ d $\left(4\frac{17}{27}\right)^{-\frac{4}{3}}$

Q5a hint $\left(\frac{64}{343}\right)^{-\frac{2}{3}} = \left(\frac{343}{64}\right)^{\frac{2}{3}}$

6 Simplify

a $\sqrt{12} + \sqrt{27}$ b $6\sqrt{2} + \sqrt{72}$

c $\sqrt{50} + \sqrt{200}$ d $\sqrt{320} - \sqrt{125}$

e $3\sqrt{112} - 2\sqrt{28}$ f $\sqrt{704} - \sqrt{99}$

Q6a hint $\sqrt{12} = \sqrt{4 \times 3} = \sqrt{4} \times \sqrt{3} = 2\sqrt{3}$

Q6 A common error is to write $\sqrt{12} + \sqrt{27} = \sqrt{39}$ or $\sqrt{320} - \sqrt{125} = \sqrt{195}$

7 Work out the value of

a $\sqrt{12} \times \sqrt{3}$ b $\sqrt{50} \times \sqrt{2}$ c $\sqrt{45} \div \sqrt{5}$ d $\sqrt{675} \div \sqrt{3}$

8 Simplify fully.

a $\sqrt{8} \times \sqrt{18}$ b $\sqrt{12} \times \sqrt{15}$ c $\frac{10\sqrt{96}}{2\sqrt{3}}$ d $\frac{21\sqrt{72}}{6\sqrt{8}}$

9 Simplify, leaving your answer in surd form where necessary.

a $\sqrt{6} \times \sqrt{8}$ b $3\sqrt{2} \times 5\sqrt{18}$ c $2\sqrt{18} \times 3\sqrt{8}$ d $7\sqrt{2} \times 5\sqrt{10}$

10 Simplify, leaving your answer in surd form where necessary.

a $\frac{6\sqrt{15}}{\sqrt{5}}$ b $\frac{4\sqrt{125}}{\sqrt{5}}$ c $\frac{4\sqrt{30}}{2\sqrt{10}}$

d $\frac{5\sqrt{108}}{2\sqrt{3}}$ e $7\sqrt{40} \times 5\sqrt{2}$ f $\frac{30\sqrt{160}}{3\sqrt{5}}$

11 Expand and simplify.

a $(5 - \sqrt{12})(2 + \sqrt{3})$

b $(4 + \sqrt{7})^2$

c $(\sqrt{32} - \sqrt{12})(\sqrt{3} - \sqrt{2})$

Q11 hint Use **FOIL** to expand the brackets. (**F**irsts, **O**uters, **I**nners, **L**asts)

Q11a hint Simplify surds fully.
$2\sqrt{12} = 2 \times \sqrt{4 \times 3} = 2 \times \sqrt{4} \times \sqrt{3} = 2 \times 2 \times \sqrt{3} = 4\sqrt{3}$

12 Expand and simplify fully.

a $(\sqrt{2} + 1)(\sqrt{2} + 5)$ b $(3 + \sqrt{12})(4 + \sqrt{3})$

c $(\sqrt{8} - 5)^2$ d $(8 - 3\sqrt{5})^2$

e $(7 - \sqrt{2})(3 + \sqrt{32})$ f $(8 + 4\sqrt{3})(8 - 4\sqrt{3})$

13 Write each expression as an integer.

a $(3 + \sqrt{2})(3 - \sqrt{2})$ b $(7 - \sqrt{2})(14 + \sqrt{8})$

c $(\sqrt{20} - 6)(\sqrt{5} + 3)$ d $(9 - \sqrt{27})(3 + \sqrt{3})$

14 **Reasoning** Work out the value of $(\sqrt{6} + 4)(\sqrt{6} - 2)(\sqrt{6} + 1)$

15 **Problem-solving** A triangle has sides of 2 cm, $(2 + \sqrt{3})$cm and $(1 + \sqrt{12})$cm.
Is the triangle right-angled?

Q15 hint Work out which is the longest side.

16 Rationalise the denominator and simplify fully.

a $\frac{14}{\sqrt{2}}$ b $\frac{18}{\sqrt{3}}$ c $\frac{\sqrt{5}}{\sqrt{2}}$

d $\frac{12\sqrt{2}}{\sqrt{6}}$ e $\frac{30}{4\sqrt{5}}$ f $\frac{5}{2\sqrt{75}}$

Q16a hint $\frac{14}{\sqrt{2}} \times \frac{\sqrt{2}}{\sqrt{2}} = \frac{\square}{\square}$

17 Write $\frac{1}{5\sqrt{48}} + \frac{1}{4\sqrt{3}}$ as a single fraction with a rational denominator.

Q17 hint Simplify $\sqrt{48}$ first.

18 Rationalise the denominator and simplify fully.

a $\frac{3}{1+\sqrt{2}}$
b $\frac{4}{\sqrt{5}-2}$
c $\frac{10}{\sqrt{5}+\sqrt{3}}$
d $\frac{14}{3+\sqrt{2}}$
e $\frac{9}{5+\sqrt{7}}$
f $\frac{22}{6-\sqrt{3}}$

1.4 Mixed exercise

Objective

- Consolidate your learning with more practice.

1 Write the upper and lower bounds for each of these measurements.

a 21.4 seconds (to 1 d.p.)
b 35 cm (to the nearest 5 cm)
c 1430 kg (to 3 s.f.)
d 16.28 litres (to 2 d.p.)

Q1 hint If the question mentions 'bounds' then you need to think about accuracy and upper and lower bound approaches.

2 **Problem-solving** A parallelogram has an area of 72 cm^2 to the nearest whole number. Its height is 6.8 cm (to 1 d.p.).
Work out the maximum length of the base of the parallelogram.
Give your answer correct to 2 d.p.

3 **Problem-solving** You can use the digits 1, 3, 4, 6 and 8.

a How many 3-digit even numbers can be made if repetitions are not allowed?
b How many 3-digit odd numbers greater than 500 can be made?

4 Work out the value of

a $49^{\frac{1}{2}}$
b $1000^{-\frac{1}{3}}$
c $16^{-\frac{1}{4}} \times 8^{\frac{2}{3}}$

5 Simplify

a $\sqrt{18}+\sqrt{98}-\sqrt{32}$
b $4\sqrt{6} \times 5\sqrt{24}$
c $\frac{10\sqrt{2}}{\sqrt{256}}$

6 **Reasoning** $x = 9.2$, $y = 8.6$ and $z = 5.4$, all correct to 1 d.p.
Work out the minimum and maximum values of $\frac{x^2}{3y - z}$
Give your answers correct to 2 d.p.

Q6 hint Show the upper and lower bounds to gain some marks even if the bounds are subsequently applied incorrectly.

7 **Problem-solving** A triangle has a base of $(6+\sqrt{18})$ cm and a height of $(12-\sqrt{72})$ cm.
Work out the area of the triangle.

8 Work out the value of

a $64^{\frac{5}{6}}$
b $36^{1.5}$
c $125^{-\frac{4}{3}}$

9 **Reasoning** Solve the equation $(9+\sqrt{5})x = 76$

10 Rationalise the denominator and simplify fully.

a $\frac{24}{5\sqrt{3}}$
b $\frac{19}{8-\sqrt{7}}$

11 Work out the value of

a $\left(\frac{216}{125}\right)^{\frac{2}{3}}$
b $\left(1\frac{15}{49}\right)^{-1.5}$
c $\left(11\frac{1}{9}\right)^{-2.5}$
d $\left(\frac{128}{2187}\right)^{-\frac{3}{7}}$

2 ALGEBRA

What is covered in this section	Higher Student Book reference
2.1 Parallel and perpendicular lines • Identify parallel and perpendicular lines. • Work out the equation of a line, given two points, or given one point and the gradient.	Unit 6
2.2 Quadratic functions • Identify and interpret quadratic functions. • Identify turning points of quadratic graphs. • Solve quadratic equations.	Units 2, 6, 9, 15
2.3 Composite and inverse functions • Use function notation. • Work out and use composite functions. • Find the inverse of a linear function.	Unit 17
2.4 Gradient and area under linear and non-linear graphs • Interpret and estimate the gradient for linear and non-linear graphs. • Interpret, calculate and estimate the area under linear and non-linear graphs.	Unit 19
2.5 The circle, $x^2 + y^2 = r^2$ • Recognise and use the equation of a circle with its centre at the origin. • Work out the gradient of the radius to the point of contact of a tangent to a circle. • Work out the equation of the tangent to a circle at a given point.	Units 6, 16
2.6 Other non-linear graphs • Recognise, sketch and interpret cubic, reciprocal, exponential and trigonometric graphs.	Units 6, 13, 15, 19
2.7 Transformations of graphs • Apply the transformations of translation and reflection to linear, quadratic, cubic, and sine and cosine functions. • Interpret transformations of graphs and write the functions algebraically.	Units 13, 19
2.8 Simultaneous equations • Solve simultaneously one linear and one quadratic equation.	Units 9, 15
2.9 Equations and expressions • Factorise quadratic expressions of the form $ax^2 + bx + c$. • Expand products of three binomial expressions. • Simplify algebraic expressions involving algebraic fractions. • Solve equations involving algebraic fractions. • Solve quadratic inequalities. • Find approximate solutions to quadratic equations numerically, using iteration.	Units 15, 17
2.10 Sequences • Recognise and use Fibonacci-type sequences, geometric sequences and quadratic sequences. • Deduce an expression for the nth term of a quadratic sequence.	Unit 2
2.11 Mixed exercise • Consolidate your learning with more practice.	

2.1 Parallel and perpendicular lines

Objectives

- Identify parallel and perpendicular lines.
- Work out the equation of a line, given two points, or given one point and the gradient.

Key point 1

The **gradient** of a line segment joining (x_1, y_1) to (x_2, y_2) is given by $\frac{y_2 - y_1}{x_2 - x_1}$.

Key point 2

The **equation of a straight line** can be written as $y = mx + c$ where m is the gradient and c is the y-intercept.

Key point 3

When two lines are **parallel**, they have the same gradient (m).
When two lines are **perpendicular**, the product of their gradients is −1.
When a graph has gradient m, a graph perpendicular to it has gradient $\frac{-1}{m}$

Warm up

1 Work out the gradient of the line segment joining each of these pairs of points.
 a (−3, 1) and (2, 6) b (4, −3) and (−6, 2) c (1, 10) and (−3, −4)

2 A straight line passes through (3, 4) and (6, −2).
 a Work out the gradient of the line.
 b Work out the equation of the line.

Q2b hint Use $y = mx + c$ to find c, using the coordinates of one of the given points and your value of m.

3 Here are the equations of six straight lines.
 A $y = 4x - 7$ B $y = 5 - 2x$ C $2x + 1 = 3y$
 D $3y + 6x = 1$ E $4 - 2y = 3x$ F $4y + 3 = x$
 a Which of these lines are parallel to each other?
 b Which of these lines are perpendicular to each other?

Q3 hint Rearrange each equation into the form $y = mx + c$.

4 **Problem-solving** A is the point (3, −2) and B is the point (5, 4).
 a Work out the gradient of the line segment AB.
 b Write down the gradient of a line perpendicular to AB.
 c Work out the equation of a line perpendicular to AB that passes through the point (−3, 2).

5 **Problem-solving** Work out the equation of the line that is perpendicular to the line $2y + 3x = 8$ and passes through the point (6, −1).

Q5 A common error is to make a mistake finding the gradient of the perpendicular line.

6 **Problem-solving** The diagram shows two lines, A and B, which intersect on the y-axis.
The equation of line A is $5x + 2y + 12 = 0$
Line B is perpendicular to line A.

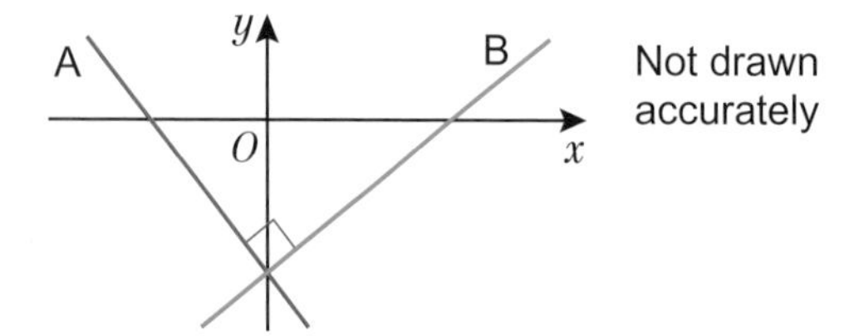

Work out the coordinates of the point of intersection of line B and the x-axis.

2.2 Quadratic functions

Objectives

- Identify and interpret quadratic functions.
- Identify turning points of quadratic graphs.
- Solve quadratic equations.

Key point 4

A **quadratic expression** contains a term in x^2 but no higher power of x. It is usually written in the form

$ax^2 + bx + c$

where a, b and c are constants and $a \neq 0$.

Key point 5

A quadratic graph has one **turning point**, which is either a minimum point or a maximum point.

maximum

minimum

Key point 6

A **quadratic equation** can have 0, 1 or 2 roots, which means that its graph crosses the x-axis at 0, 1 or 2 points.

Key point 7

$x^2 + bx + c$ can be written in the form $\left(x + \frac{b}{2}\right)^2 - \left(\frac{b}{2}\right)^2 + c$.

This is called **completing the square**.
Completing the square identifies the line of symmetry and the turning point of a quadratic graph.

Key point 8

You can use the **quadratic formula** $x = \frac{-b \pm \sqrt{b^2 - 4ac}}{2a}$

to find the solution to a quadratic equation $ax^2 + bx + c = 0$

Key point 9

When you are asked to give your answers to a *required degree of accuracy* (e.g. 2 d.p.), you will need to use the quadratic formula.

When you are asked to give *exact answers* (which usually implies answers in surd form), or to identify turning points, you will need to solve by completing the square. Answers should be written in the form $x = m \pm \sqrt{n}$.

Key point 10

In the quadratic formula, the expression under the square root sign ($b^2 - 4ac$) is called the **discriminant**. It tells you whether the equation has 0, 1 or 2 real roots.

- When $b^2 - 4ac > 0$ there are two distinct solutions to the quadratic equation (two real roots).
- When $b^2 - 4ac = 0$ there is one solution to the quadratic equation (one real root).
- When $b^2 - 4ac < 0$ there are no real solutions to the quadratic equation (no real roots).

Warm up

1 Here are the graphs of four quadratic functions.

a

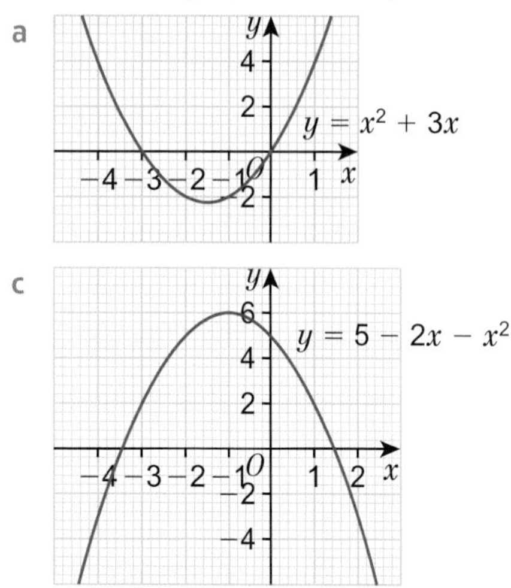

b

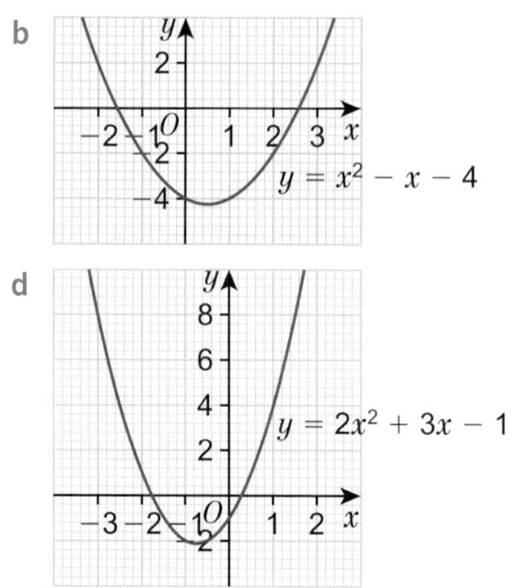

c $y = 5 - 2x - x^2$

d $y = 2x^2 + 3x - 1$

For each of these quadratic graphs

i write down the equation of the line of symmetry

ii write down the coordinates of the turning point, stating whether it is a maximum point or a minimum point

iii write down the roots of the corresponding quadratic equation.

2 Problem-solving

a Copy and complete the table of values for $y = x^2 - x - 3$

x	−3	−2	−1	0	1	2	3	4
y			−1					9

b Draw the graph of $y = x^2 - x - 3$ on a suitable grid.

c Use your graph to solve the equation $x^2 - x - 3 = 0$

d Solve the equation $x^2 - 2x - 4 = 0$

Q2 A common error is is to not use the graph from part **b** to answer parts **c** and **d**.

3 Problem-solving

a Copy and complete the table of values for $y = 2x^2 - 5x + 1$

x	−1	0	1	2	3	4
y					4	

b Draw the graph of $y = 2x^2 - 5x + 1$ on a suitable grid.

c Use your graph to solve the equation $2x^2 - 5x + 1 = 0$

d Solve the equation $2x^2 - 3x - 3 = 0$

4 Expand and simplify

a $(x + 3)(x - 8)$ b $(2x - 5)(x - 4)$ c $(5x + 3)(2x + 5)$

d $(3x + 7)(2x - 1)$ e $(x + 2y)(3x - 4y)$ f $(4x - 3y)(3x - 2y)$

5 Factorise

a $x^2 - 9$ b $36 - 25y^2$

c $5x^2 - 20$ d $12 - 27w^2$

e $45c^2 - 20d^2$ f $32m^2 - 50t^2$

Q5 hint All these expressions are 'a difference of two squares'.
$a^2 - b^2 = (a + b)(a - b)$
Some have a common factor. Be careful.

6 Solve

a $(2x - 3)(x + 5) = 0$ b $2x^2 - 13x + 15 = 0$ c $2x^2 - x - 28 = 0$

d $5x^2 + 38x + 21 = 0$ e $15x = 4 - 4x^2$ f $4x^2 = 27 - 12x$

7 Solve, giving your answers to 2 d.p.

a $x^2 - 4x + 2 = 0$ b $y^2 + 4y - 10 = 0$

c $2w^2 - w - 5 = 0$ d $3x^2 - 7x = 1$

e $5y^2 - 2y = 5$ f $4w^2 + 1 + 6w = 0$

Q7 hint Use the quadratic formula,

$$x = \frac{-b \pm \sqrt{b^2 - 4ac}}{2a}$$

8 **Reasoning** Decide whether each of these quadratic equations has 0, 1 or 2 real roots.

a $x^2 + 2x - 7 = 0$ b $y^2 - 3y + 4 = 0$ c $2w^2 - 5w - 6 = 0$

d $x^2 - 14x + 49 = 0$ e $3y^2 = 11 - 5y$ f $5w^2 = 6w - 2$

9 **Problem-solving** The quadratic equation $x^2 + mx + 64 = 0$ has exactly one solution. Work out two possible values for m.

10 **Problem-solving** For what value of k does the equation $5x^2 - 8x + k = 0$ have exactly one solution?

11 Write each of these expressions in the form

$a(x + b)^2 + c$

where a, b and c are integers.

a $x^2 + 2x + 10$ b $x^2 - 10x + 30$

c $x^2 + 8x - 1$ d $2x^2 + 4x - 3$

e $3x^2 - 6x + 1$ f $4x^2 - 16x + 5$

Q11d hint Remove a common factor from the first two terms.

$2x^2 + 4x - 3 = 2(x^2 + 2x) - 3 = 2[(x + 1)^2 - \square] - 3$

12 Solve these by completing the square, giving your answers in surd form.

a $x^2 + 2x - 7 = 0$ b $x^2 + 12x - 4 = 0$ c $x^2 - 8x + 13 = 0$

d $x^2 + 6x + 2 = 0$ e $x^2 - 10x + 20 = 0$ f $4x^2 - 20x + 1 = 0$

13 By completing the square, identify the turning points of the following graphs. Sketch each graph.

a $y = x^2 - 6x + 4$ b $y = 2x^2 + 4x - 5$ c $y = 3x^2 - 12x + 7$

14 **Reasoning** By completing the square, show that the equation $2x^2 - 6x + 7 = 0$ has no real roots.

2.3 Composite and inverse functions

Objectives

- Use function notation.
- Work out and use composite functions.
- Find the inverse of a linear function.

Key point 11

A **function** is a rule for working out values of y for given values of x.
For example, $y = 3x$ and $y = x^2$ are functions.
The notation **f(x)** is read as 'f of x'. f is the function. $f(x) = 3x$ means the function of x is $3x$.

Key point 12

fg is a **composite function**. To work out fg(x), first work out g(x) and then substitute your answer into f(x).

Key point 13

The **inverse function** reverses the effect of the original function.
$f^{-1}(x)$ is the inverse of f(x).

Warm up

1 $f(x) = 1 - 3x^2$

Work out

a f(0) b f(2) c f(−4) d $f(\frac{1}{2})$

2 $g(x) = 4x - 7$

Work out the value of n when

a $g(n) = 1$ b $g(n) = -5$ c $g(n) = 0$ d $g(n) = 11$

3 $h(x) = x^2 - 2x$

Work out the values of t when

a $h(t) = 3$ b $h(t) = 15$ c $h(t) = 80$

Q3 hint Set up a quadratic equation, rearrange and solve.

4 $f(x) = 6x + 5$

Work out, simplifying fully

a $2f(x) + 6$ b $4f(2x)$ c $f(3x) - 7$

Q4b hint Replace x with $2x$.

Q4c hint Replace x with $3x$.

5 $f(x) = 2x^2 + 3x$ $g(x) = (x - 4)(x + 1)$

Work out the values of m when

a $f(m) = -1$ b $f(m) = 5$ c $g(m) = 6$ d $g(m) = 24$

6 Reasoning $g(x) = 5x^2 + 3$

Work out, simplifying fully

a $g(-x)$ b $g(2x)$ c $g(x + 2)$ d $g(3x - 4)$

7 Reasoning $f(x) = x^2 - 5$ $g(x) = 1 - 4x$

Work out

a $gf(3)$ b $fg(3)$ c $gf(-2)$

d $fg(-2)$ e $gf(x)$ f $fg(x)$

Q7a hint $gf(x)$ means use $f(x)$ first, then substitute your answer into $g(x)$.

Q7e, f Simplify fully.

8 Reasoning $f(x) = 7 - x$ $g(x) = 5x - 2$ $h(x) = x^2 + 6$

Work out, simplifying fully

a $gf(x)$ b $fg(x)$ c $fh(x)$

d $hf(x)$ e $gh(x)$ f $hg(x)$

Q8 A common error with composite functions is to apply the 'outside' function first.

9 Reasoning $f(x) = 2x - 5$ $g(x) = (x + 6)(x - 1)$

a Work out $gf(x)$, simplifying fully.

b Solve the equation $gf(x) = 0$

10 Reasoning Work out the inverse, $f^{-1}(x)$, of each of these linear functions.

a $f(x) = 3x - 4$ b $f(x) = 3(x - 4)$ c $f(x) = \frac{1}{2}x + 5$

d $f(x) = \frac{1}{2}(x + 5)$ e $f(x) = 2(x - 6) - 3$ f $f(x) = 16 - 3(x + 2)$

Q10 hint You could use a function machine to help.

2.4 Gradient and area under linear and non-linear graphs

Objectives

- Interpret and estimate the gradient for linear and non-linear graphs.
- Interpret, calculate and estimate the area under linear and non-linear graphs.

Key point 14

When the equation of a linear graph is written in the form $y = mx + c$, m represents the gradient. The gradient is also known as the **rate of change**.

Key point 15

On a **distance–time graph**, the gradient is the rate of change of distance over time, or the speed.

The average speed is the total distance travelled divided by the total time taken. $s = \frac{d}{t}$

For a non-linear distance–time graph, the gradient of the tangent at any point gives the speed at that point.

Key point 16

On a **velocity–time graph**, the gradient is the rate of change of velocity over time, or the acceleration.

$a = \frac{v}{t}$

The area under a velocity–time graph represents the distance travelled.

Key point 17

The **tangent** to a curved graph is a straight line that touches the graph at a point.
The gradient at a point on a curve is the gradient of the tangent at that point.

Key point 18

The straight line that connects two points on a curve is called a **chord**.
The gradient of the chord gives the average rate of change and can be used to find the average speed on a distance–time graph.

The area under a velocity–time graph shows the displacement, or distance from the starting point. To estimate the area under a part of a curved graph, draw a chord between the two points you are interested in, and straight lines down to the horizontal axis to create a trapezium. The area of the trapezium is an estimate for the area under this part of the graph.

1 The distance–time graph represents train journeys between Newcastle and London.

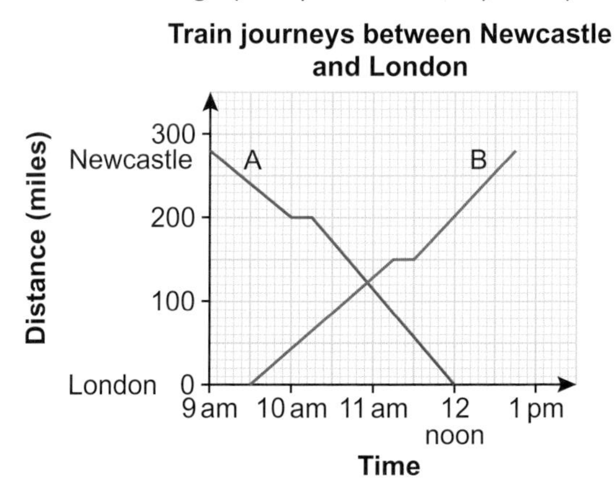

a Estimate how far from London the trains are when they pass each other.

b Work out the speed for each part of the journey for train A.

c Work out the average speed for the journey for train B.

2 The velocity–time graph describes the motion of a particle over a 10-second period during an experiment in a laboratory.

a What was the maximum velocity of the particle?

b For how many seconds was the particle moving at a constant speed?

c Work out the particle's acceleration during the first 2 seconds of its motion.

d Work out the distance travelled by the particle in this 10-second period.

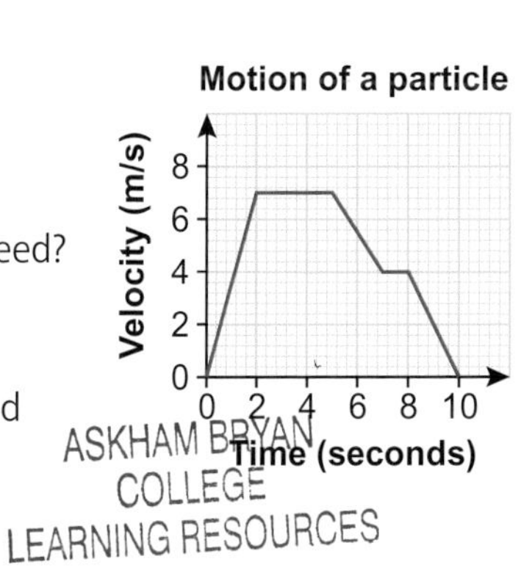

Warm up

3 The velocity–time graph describes the motion of a car over a 20-second period.

- **a** Work out the deceleration of the car between 13 and 15 seconds after the start.
- **b** Describe fully the motion of the car during this 20-second period.
  Give details of speed, acceleration and deceleration and times for each part of the graph.
- **c** How far did the car travel in this 20-second period?
- **d** Work out the average speed of the car over the 20-second period.

Motion of a car

Velocity (m/s)

Time (seconds)

4 This is the graph of $y = x^2 - x - 4$

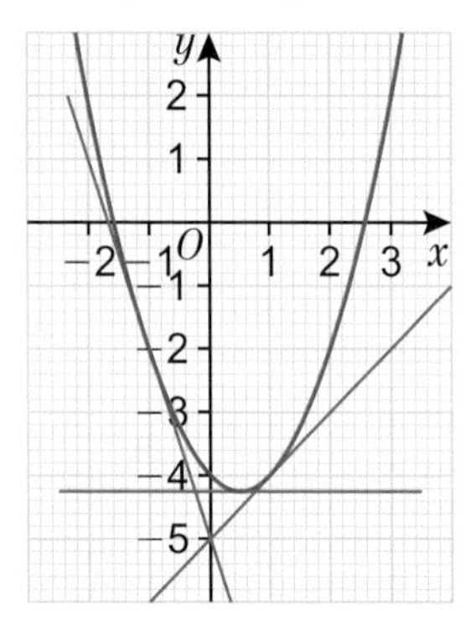

> **Q4 hint** The graph has a turning point at $x = 0.5$.
> The graph has negative gradient when $x < 0.5$.
> The graph has positive gradient when $x > 0.5$.

> **Q4 hint** To work out the gradient, choose two points on the tangent and form a right-angled triangle.
> Calculate $\frac{\text{change in } y}{\text{change in } x}$

Tangents have been drawn at $x = -1$, $x = 0.5$ and $x = 1$.
Work out the gradient of the graph at each of these points.

5 **a** Copy and complete the table of values for $y = x^2 + 5x + 1$

x	−6	−5	−4	−3	−2	−1	0	1
y	7				−5			

- **b** Draw the graph of $y = x^2 + 5x + 1$
- **c** Work out the gradient of the graph when $x = -2$.
- **d** Work out the gradient of the graph when $x = -4$.

> **Q5c, d hint** Draw your tangents so that they just touch the curve at the given points. Place your ruler on the outside of the curve and judge 'by eye' where to draw the line. Draw it a reasonable length and then choose two points on it and work out the gradient.

6 **a** Copy and complete the table of values for $y = 3 + 4x - x^2$

x	−1	0	1	2	3	4	5
y			6				−2

- **b** Draw the graph of $y = 3 + 4x - x^2$
- **c** Work out the gradient of the graph when $x = 1$.
- **d** Work out the gradient of the graph when $x = 4$.

7 **a** Copy and complete the table of values for $y = 2x^2 + 7x - 2$

x	−4	−3	−2	−1	0	1
y	2			−7		

- **b** Draw the graph of $y = 2x^2 + 7x - 2$
- **c** Work out the gradient of the graph when $x = -2.5$.
- **d** Work out the gradient of the graph when $x = -0.5$.

8 **Reasoning** This is the graph of $y = 8 + 2x - x^2$

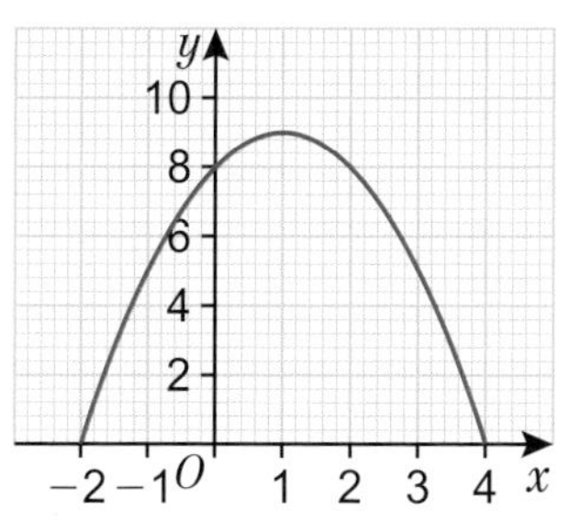

The section between $x = -1$ and $x = 0$ looks like this, with a chord drawn between (−1, 5) and (0, 8).

a Work out the shaded area of this section.

b Work out the areas of the trapezia and triangles formed by joining successive points with chords, as in this diagram.

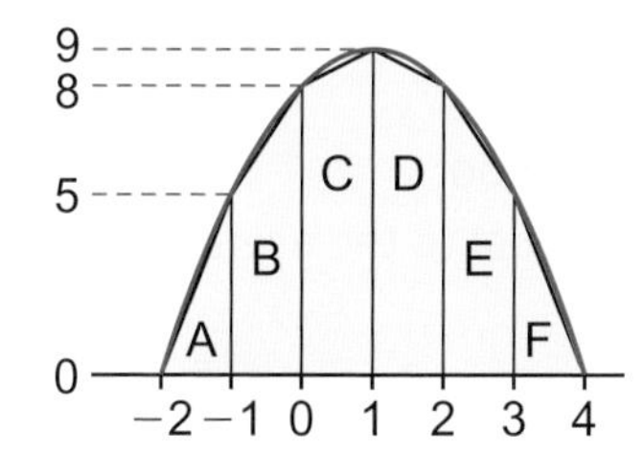

> **Q8a hint** The formula for the area of a trapezium is $A = \frac{1}{2}(a + b)h$.
> So the area of the shaded region is $\frac{1}{2} \times (5 + 8) \times 1 = \square$ units of area

> **Q8b hint** Section A is triangle so Area A = $\frac{1}{2} \times 1 \times 5 = \square$ units of area

c What is the sum of the areas of sections A to F?

d Explain why your answer to part **c** is a good approximation to the area under the graph of $y = 8 + 2x - x^2$ from $x = -2$ to $x = 4$.

9 **Reasoning**

a Copy and complete the table of values for $y = x^2 + x$

x	−2	−1	0	1	2	3
y					6	

b Draw the graph of $y = x^2 + x$

c Estimate the area under the graph of $y = x^2 + x$ from $x = 0$ to $x = 3$.

d Is your answer to part **c** an underestimate or an overestimate of the actual area? Justify your answer.

10 **Reasoning**

a Draw the graph of $y = 7 + 6x - x^2$ for $-1 \leqslant x \leqslant 7$.

b Estimate the area enclosed by the graph and the x-axis.

11 **Reasoning** The velocity–time graph describes the motion of a train leaving a station.
The train travels in a straight line.

a Work out the average acceleration from time $t = 50$ to $t = 100$ seconds.

b Estimate the acceleration at time $t = 150$ seconds.

c Estimate the distance travelled in 250 seconds.

Motion of a train

12 Problem-solving The velocity–time graph gives some information about the motion of a car. The car travels in a straight line.

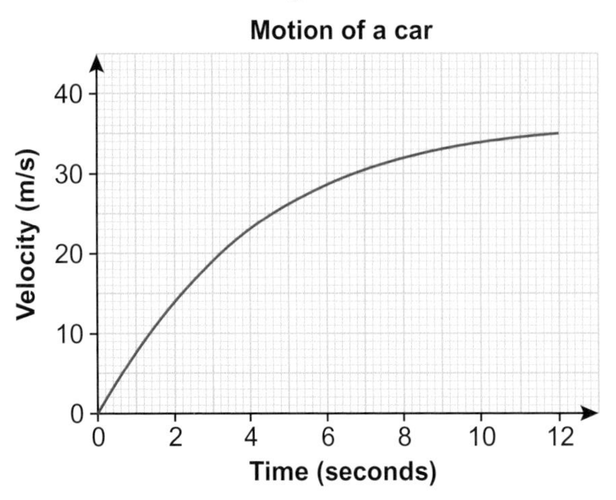

a Estimate the acceleration at time $t = 2$ seconds.

b Estimate the total distance travelled in the 12-second period.

c The instantaneous acceleration at a time T seconds is equal to the average acceleration over the 12-second period.
Work out an estimate for the value of T.

2.5 The circle, $x^2 + y^2 = r^2$

Objectives

- Recognise and use the equation of a circle with its centre at the origin.
- Work out the gradient of the radius to the point of contact of a tangent to a circle.
- Work out the equation of the tangent to a circle at a given point.

Key point 19

The **equation of a circle** with centre (0, 0) and radius r is $x^2 + y^2 = r^2$.

Key point 20

A **tangent** is a straight line that touches a circle at one point only.
The angle between a tangent and the radius is 90°.

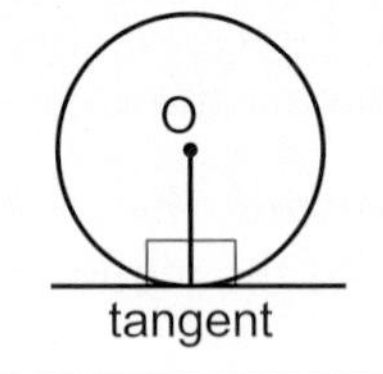

Warm up

1 Which of the following represent the equation of a circle?
Justify your answer by stating the radius of the circle.

a $x + y = 25$ b $x^2 + y^2 = 4$ c $x^2 - y^2 = 25$

d $x^2 + 2y^2 = 10$ e $y^2 = 81 - x^2$ f $x^2 + y^2 = 6$

Q1 hint Compare each equation to $x^2 + y^2 = r^2$.

2 Each value for m represents the gradient of a line.
For each one, write down the gradient of a line that is perpendicular to the given line.

a $m = 3$ b $m = -4$ c $m = \frac{1}{2}$ d $m = -\frac{1}{5}$

e $m = \frac{4}{5}$ f $m = 3\frac{1}{2}$ g $m = -2\frac{1}{2}$ h $m = \frac{4}{19}$

3 a Draw the graphs of $x^2 + y^2 = 9$ and $2y - 5x = 10$ on the same axes.
b Write down the coordinates of the points where these graphs intersect.

Q3 hint Use the same scale on both axes.

Q3 A common error is to not recognise that a graph in the form $x^2 + y^2 = r^2$ is a circle.

4 a Draw the graphs of $x^2 + y^2 = 16$ and $y = x^2 - 3$ on the same axes.
b Write down the coordinates of the points where these graphs intersect.

5 **Reasoning** A circle has equation $x^2 + y^2 = 25$
a Write down the radius of the circle.
b Show that the point (3, 4) lies on the circumference of the circle.
c Work out the gradient of the radius joining the centre of the circle to the point (3, 4).
d What is the gradient of the tangent to the circle at the point (3, 4)?
e Work out the equation of the tangent at the point (3, 4).

Q5e hint Use $y = mx + c$. Your answer to part **d** is the value of m and you know that the tangent passes through (3, 4). Substitute these values into $y = mx + c$ to find c.

6 **Problem-solving** A circle has equation $x^2 + y^2 = 20$
a Write down the radius of the circle.
b Work out the equation of the tangent to the circle at the point (−4, 2).

7 **Problem-solving** A circle has equation $x^2 + y^2 = 45$
a Its radius is $m\sqrt{n}$. State the values of m and n.
b The point (−3, p) lies on the circle. Work out the two possible values of p.
c Work out the equations of the tangents at the two points corresponding to your answers to part **b**.
d Draw a sketch graph to show the circle and the two tangents.

8 **Problem-solving** A circle has equation $x^2 + y^2 = 169$. The straight line $x = 5$ intersects the circle in two places.
a Work out the coordinates of the points of intersection.
b Work out the equations of the tangents at the two points of intersection.
c Draw a sketch graph to show the circle and the two tangents at the points where $x = 5$.
d The two tangents intersect on the x-axis at the point (p, 0). Work out the value of p.

2.6 Other non-linear graphs

Objective

- Recognise, sketch and interpret cubic, reciprocal, exponential and trigonometric graphs.

Key point 21

A **cubic function** contains a term in x^3 but no higher power of x.
It is usually written in the form $y = ax^3 + bx^2 + cx + d$, where a, b, c and d are constants and $a \neq 0$.
When $a > 0$ the function looks like

or

When $a < 0$ the function looks like

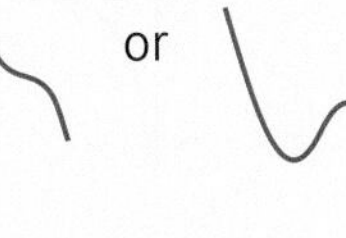

The graph intersects the y-axis at the point $y = d$.
The roots of the graph can be found by finding the values of x for which $y = 0$.

Key point 22

When the graph of a cubic function y crosses the x-axis three times, the equation $y = 0$ has three solutions.

For example $y = (x + 2)(x - 1)(x - 3)$

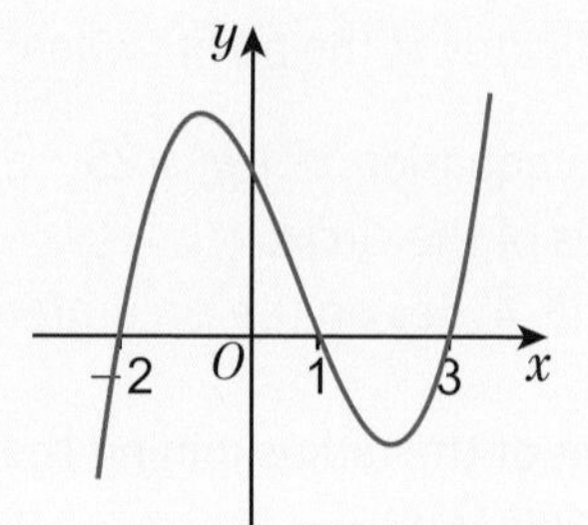

When the graph of a cubic function y crosses the x-axis and touches the x-axis once, the equations $y = 0$ has three solutions but one of them is repeated.

For example $y = (x - 1)(x + 2)^2$

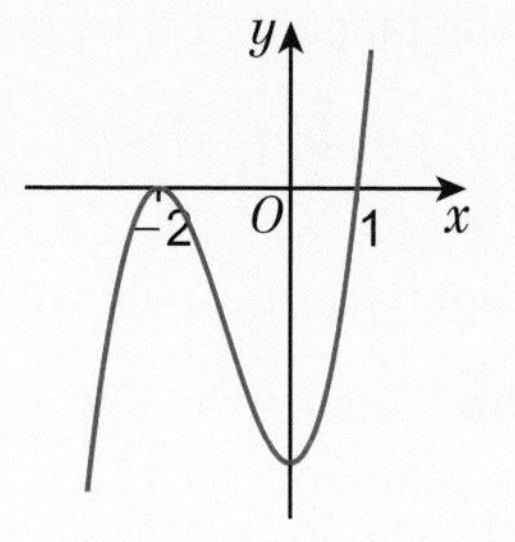

When the graph of a cubic function y crosses the x-axis once, the equation $y = 0$ can have

- one distinct, repeated solution, for example $y = (x - 1)^3$

or

- only one real solution, for example $(x + 2)(x^2 + x + 1)$
  The quadratic $(x^2 + x + 1)$ has no real solutions.

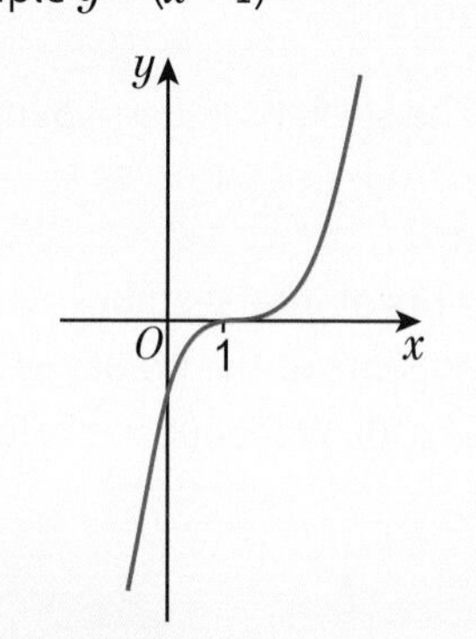

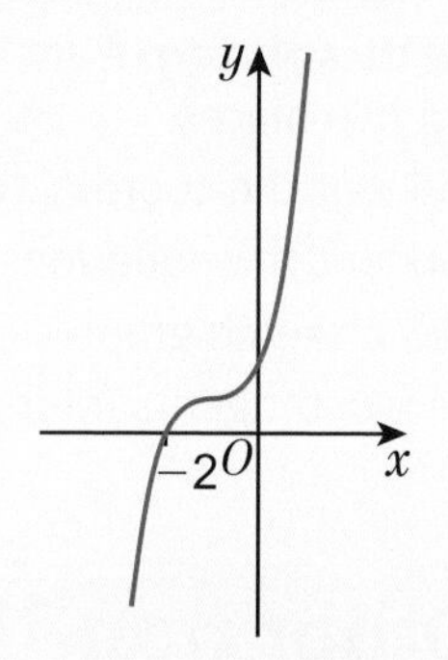

Key point 23

$y = \frac{k}{x}$ and $y = \frac{k}{x^2}$ are examples of **reciprocal functions**.

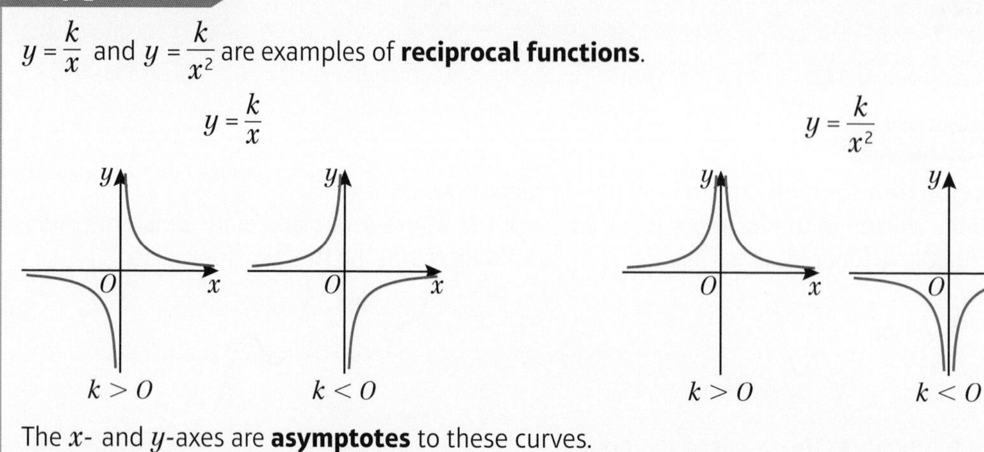

The x- and y-axes are **asymptotes** to these curves.

An asymptote is a line that a graph gets very close to but never actually touches.

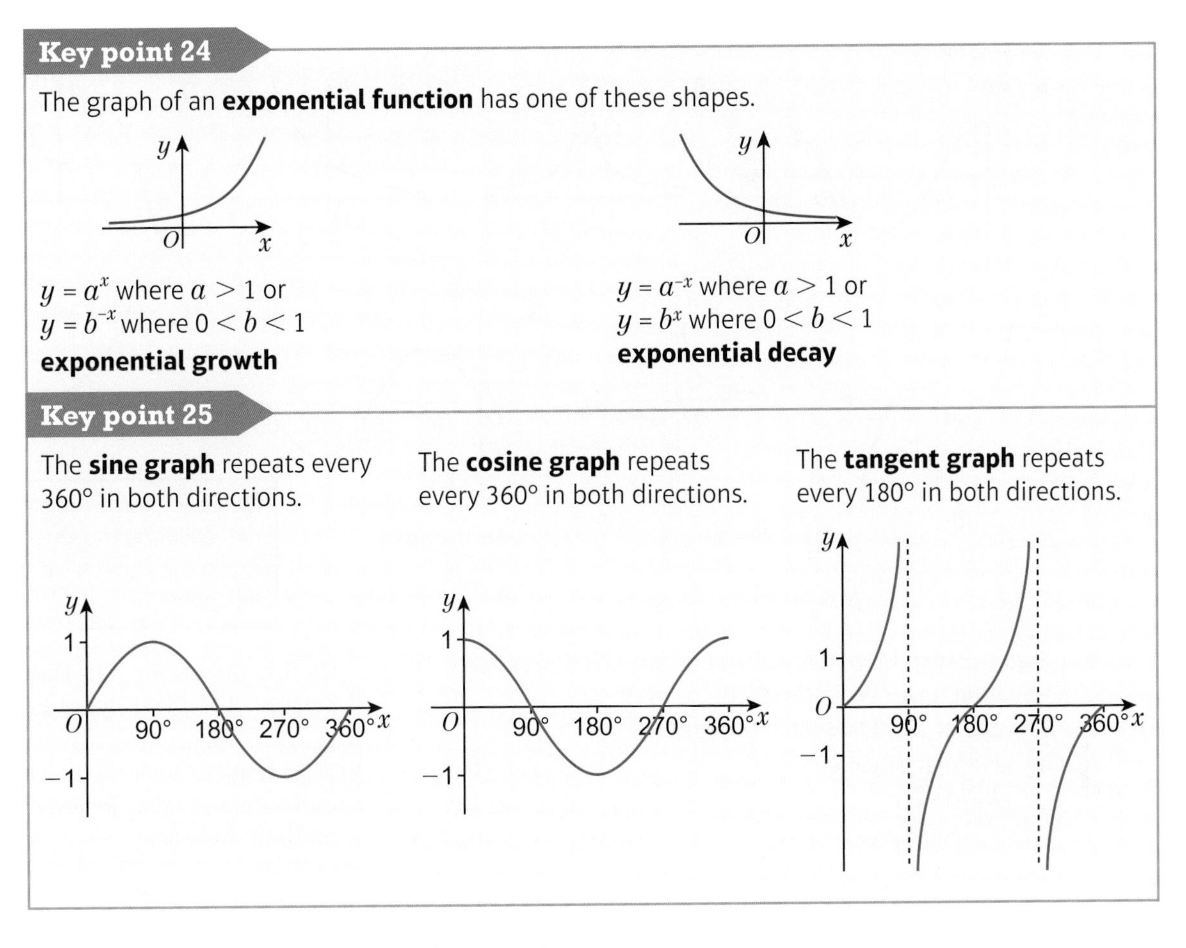

Key point 24

The graph of an **exponential function** has one of these shapes.

$y = a^x$ where $a > 1$ or
$y = b^{-x}$ where $0 < b < 1$
exponential growth

$y = a^{-x}$ where $a > 1$ or
$y = b^x$ where $0 < b < 1$
exponential decay

Key point 25

The **sine graph** repeats every 360° in both directions.

The **cosine graph** repeats every 360° in both directions.

The **tangent graph** repeats every 180° in both directions.

1 Match each equation to one of these graphs.

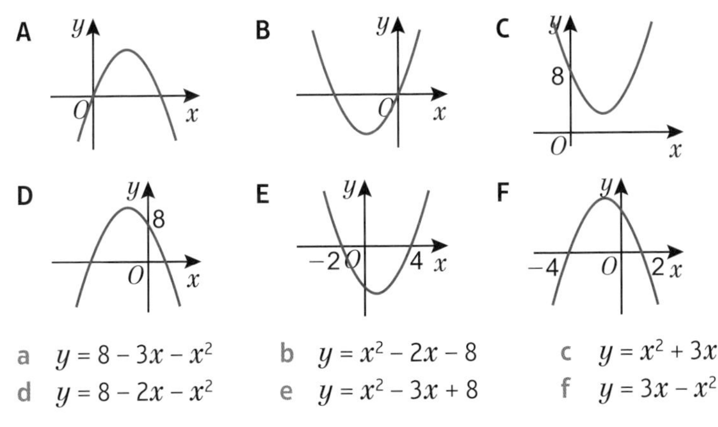

a $y = 8 - 3x - x^2$
b $y = x^2 - 2x - 8$
c $y = x^2 + 3x$
d $y = 8 - 2x - x^2$
e $y = x^2 - 3x + 8$
f $y = 3x - x^2$

2 Sketch the graphs of the reciprocal function $y = \frac{8}{x}$ and the straight line $y = \frac{1}{2}x$ on the same axes.
Work out the coordinates of the points of intersection of these graphs.

3 **Reasoning**

a Sketch the graphs of $y = \frac{6}{x}$ and $y = 4 - \frac{1}{2}x$ on the same axes.

b Show that the points of intersection are given by the solutions to the equation $x^2 - 8x + 12 = 0$

c Work out the coordinates of the points of intersection.

4 **Reasoning** Match each equation to one of these graphs.

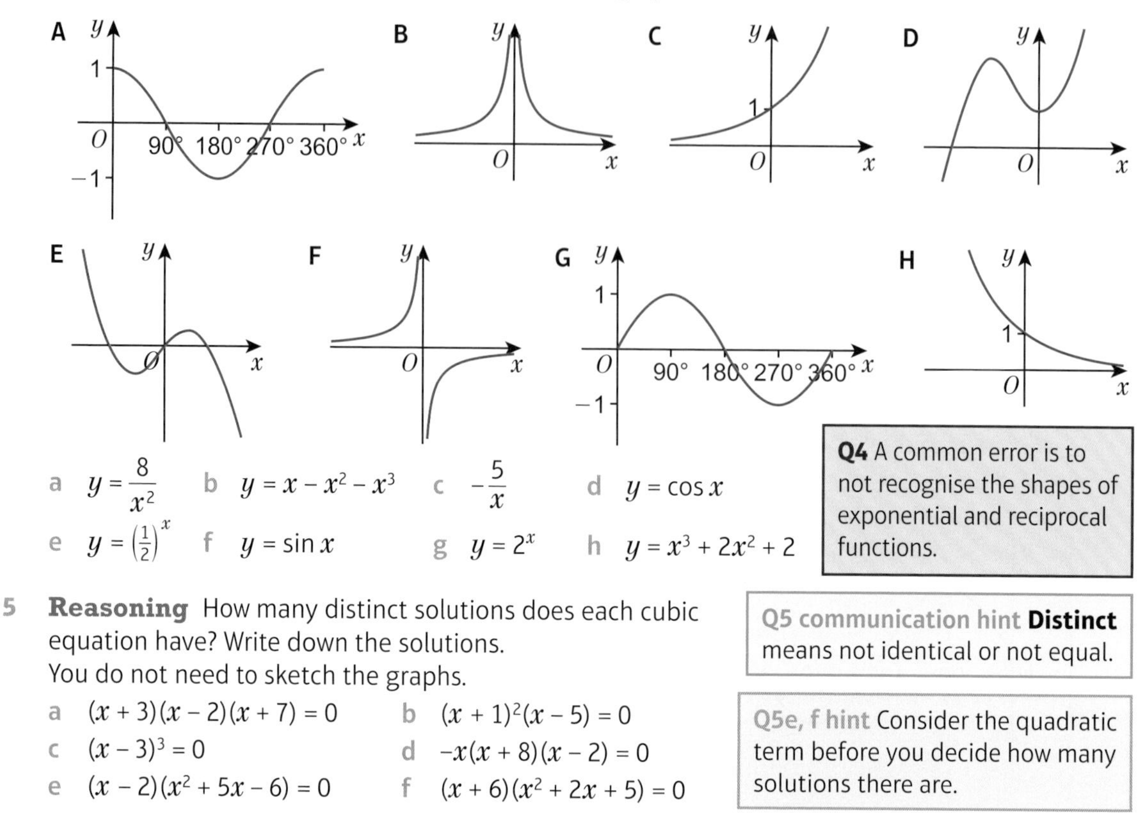

a $y = \frac{8}{x^2}$ b $y = x - x^2 - x^3$ c $-\frac{5}{x}$ d $y = \cos x$

e $y = \left(\frac{1}{2}\right)^x$ f $y = \sin x$ g $y = 2^x$ h $y = x^3 + 2x^2 + 2$

Q4 A common error is to not recognise the shapes of exponential and reciprocal functions.

5 **Reasoning** How many distinct solutions does each cubic equation have? Write down the solutions.
You do not need to sketch the graphs.

a $(x + 3)(x - 2)(x + 7) = 0$ b $(x + 1)^2(x - 5) = 0$

c $(x - 3)^3 = 0$ d $-x(x + 8)(x - 2) = 0$

e $(x - 2)(x^2 + 5x - 6) = 0$ f $(x + 6)(x^2 + 2x + 5) = 0$

Q5 communication hint Distinct means not identical or not equal.

Q5e, f hint Consider the quadratic term before you decide how many solutions there are.

6 **Reasoning** Sketch these cubic graphs, marking clearly the points of intersection with the x- and y-axes.

a $y = (x + 1)(x - 1)(x - 4)$ b $y = (x - 2)(x - 3)(x + 5)$

c $y = x(x - 5)^2$ d $y = (x - 2)^2(x + 3)$

e $y = -x^2(x + 4)$ f $y = -x^3 + 3x^2 + 10x$

Q6f hint Factorise fully first.

7 **Reasoning** This exponential graph shows how a culture of bacteria grows over time.

a How many bacteria were present at the start?

b How long did it take for the number of bacteria to reach 2000?

c Estimate the rate of increase of the number of bacteria after $3\frac{1}{2}$ hours.

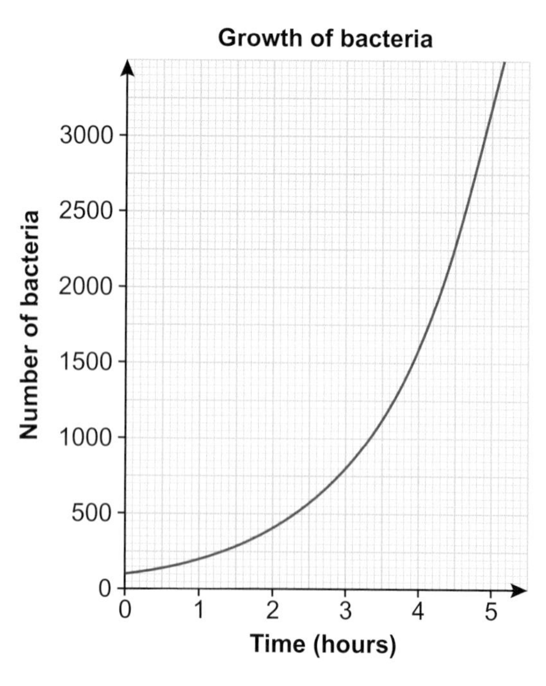

8 **Problem-solving** The graph shows the relationship between temperature and time as a cup of coffee cooled down.

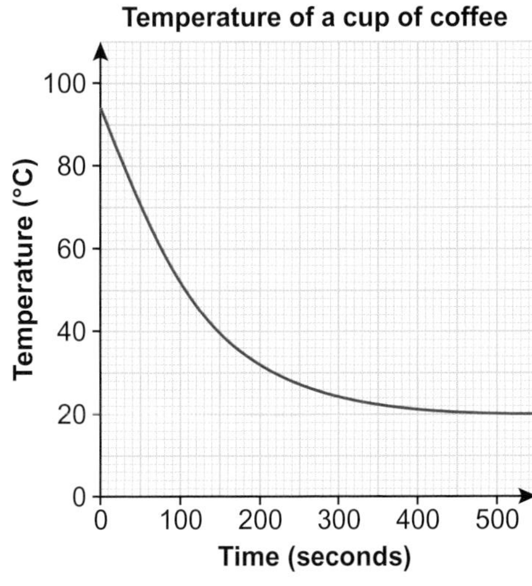

a How long did it take for the temperature to reach 50 °C?

b Calculate the average rate of fall of temperature between time t = 100 seconds and t = 200 seconds.

c Calculate the actual rate of fall of temperature after 300 seconds.

d The actual rate of fall of temperature after T seconds is equal to the average rate of fall of temperature over the whole 500-second period.
Use the graph to estimate the value of T.

9 **Reasoning** You are given that sin 54° = 0.8090

a Write down another angle in the range $0° \leqslant x \leqslant 360°$ for which $\sin x$ = 0.8090

b Solve $\sin x = -0.8090$ for $0° \leqslant x \leqslant 360°$

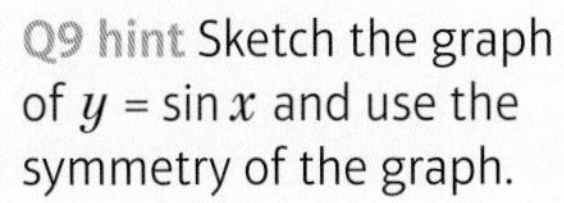
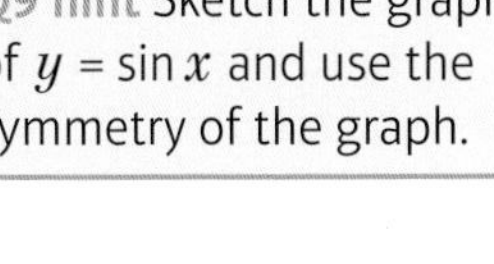

Q9 hint Sketch the graph of $y = \sin x$ and use the symmetry of the graph.

10 **Problem-solving**

a Write down the two solutions of the equation $n^2 = 0.64$

b Sketch the graph of $y = \cos x$ for $0° \leqslant x \leqslant 360°$

c Work out all the solutions of the equation $(\cos x)^2 = 0.64$ for $0° \leqslant x \leqslant 360°$
Give solutions correct to 1 d.p.

11 **Problem-solving** Solve the equation $3 \tan x = 12.6$ for $0° \leqslant x \leqslant 720°$
Give solutions correct to 1 d.p.

2.7 Transformations of graphs

Objectives

- Apply the transformations of translation and reflection to linear, quadratic, cubic, and sine and cosine functions.
- Interpret transformations of graphs and write the functions algebraically.

Key point 26

The transformation that maps the graph of $y = f(x)$ onto the graph of $\boldsymbol{y = f(x) + a}$ is a translation by $\begin{pmatrix} 0 \\ a \end{pmatrix}$.

The transformation that maps the graph of $y = f(x)$ onto the graph of $\boldsymbol{y = f(x + a)}$ is a translation by $\begin{pmatrix} -a \\ 0 \end{pmatrix}$.

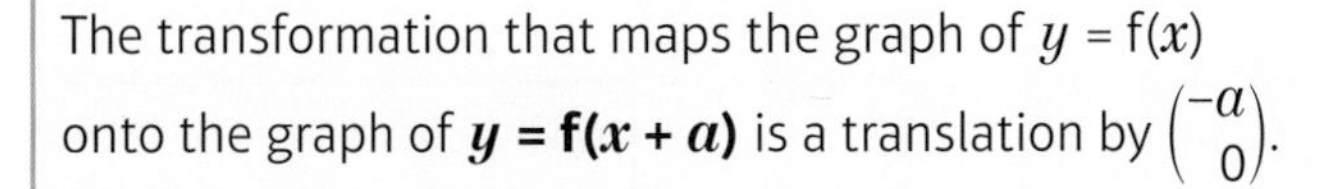

Key point 27

The transformation that maps the graph of $y = f(x)$ onto the graph of $\boldsymbol{y = f(-x)}$ is a reflection in the y-axis.

The transformation that maps the graph of $y = f(x)$ onto the graph of $\boldsymbol{y = -f(x)}$ is a reflection in the x-axis.

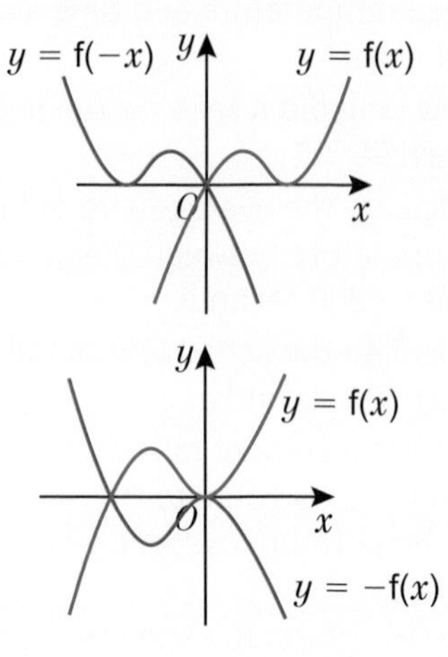

Key point 28

The transformation that maps the graph of $y = f(x)$ onto the graph of $\boldsymbol{y = af(x)}$ is a stretch of scale factor a parallel to the y-axis.

The transformation that maps the graph of $y = f(x)$ onto the graph of $\boldsymbol{y = f(ax)}$ is a stretch of scale factor $\frac{1}{a}$ parallel to the x-axis.

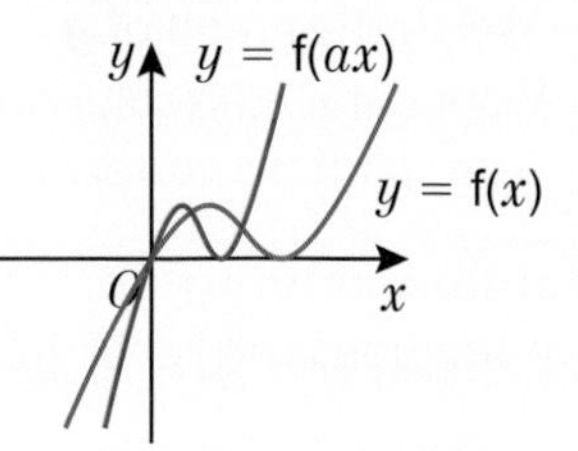

Stretches are not a requirement of the specification but are included in this section for additional challenge.

Warm up

1 This is the graph of $y = 2x + 1$

Copy the axes and the graph of $y = 2x + 1$ and, on the same axes

- **a** translate the graph by $\begin{pmatrix} 0 \\ 2 \end{pmatrix}$, and label the new graph A
- **b** translate the graph by $\begin{pmatrix} -2 \\ 0 \end{pmatrix}$, and label the new graph B
- **c** reflect the graph in the y-axis, and label the new graph C
- **d** reflect the graph in the x-axis, and label the new graph D.

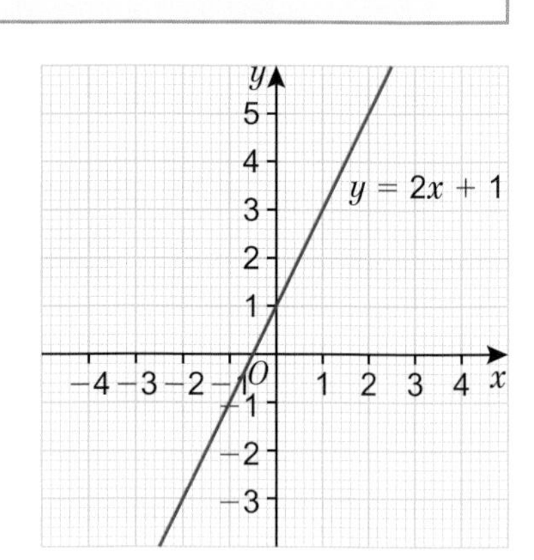

2 Work out the equations of the transformed graphs in **Q1**.

3 $f(x) = 2x + 1$

- **a** Work out, and simplify fully
  - **i** $f(x) + 2$ **ii** $f(x + 2)$ **iii** $f(-x)$ **iv** $-f(x)$
- **b** What do you notice about your answers to **Q2** and **Q3**?

4 **Reasoning** $f(x) = 4 - 2x$

- **a** Draw the graph of $y = f(x) = 4 - 2x$
- **b** On the same axes, draw the graphs of
  - **i** $y = f(x) - 4$ **ii** $y = f(x + 3)$ **iii** $y = f(-x)$
  - **iv** $y = -f(x)$ **v** $y = 2f(x)$ **vi** $y = f(2x)$
- **c** Describe fully each transformation in part **b**.
- **d** Work out the algebraic equation of each transformed graph. Simplify fully.

> **Q4b** A common error in part **ii** is to draw the graph of $y = f(x - 3)$ instead of $y = f(x + 3)$.

5 **Reasoning** This is the graph of $y = f(x) = x^2 - 2$

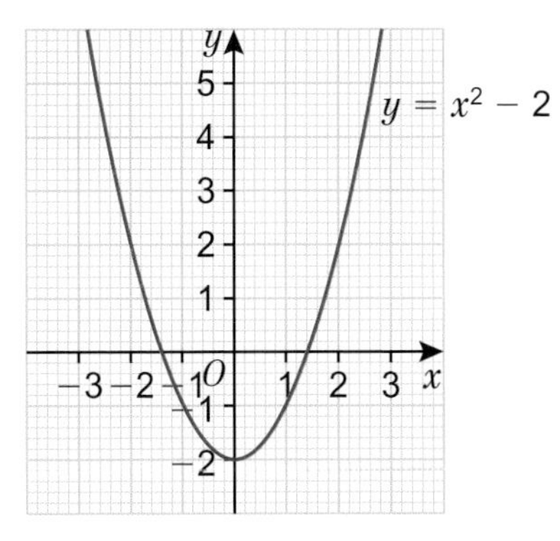

a Copy the diagram and, on the same axes, draw the graphs of
  i $y = f(x) + 3$ ii $y = f(x + 1)$ iii $y = -f(x)$ iv $y = f(2x)$

b Describe fully each transformation in part **a**.

c Work out the algebraic equation of each transformed graph. Simplify fully.

6 **Reasoning** This is the graph of $y = f(x) = 4 - x^2$

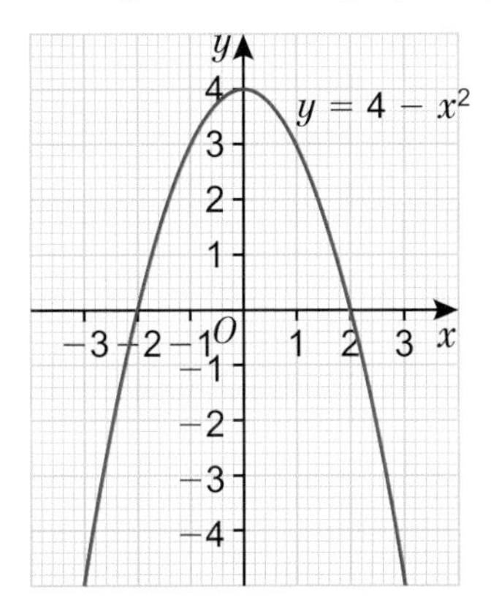

a Copy the diagram and, on the same axes, draw the graphs of
  i $y = 2f(x)$ ii $y = f(2x)$ iii $y = \frac{1}{2}f(x)$ iv $y = f\left(\frac{1}{2}x\right)$

b Describe fully each transformation in part **a**.

c Work out the algebraic equation of each transformed graph. Simplify fully.

7 **Reasoning** This is the graph of $y = f(x) = x^3 + 2x^2$

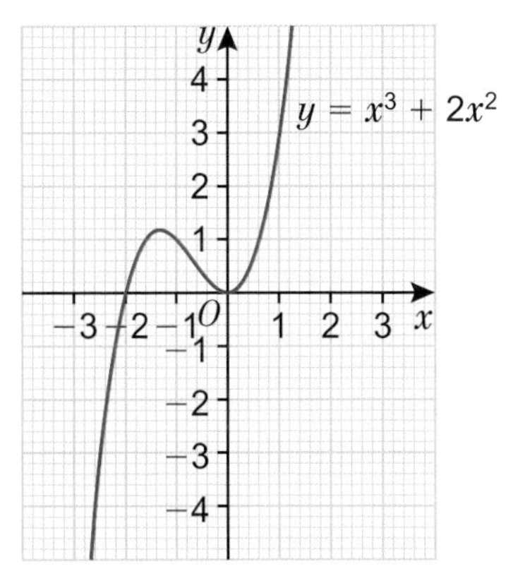

a Copy the diagram and, on the same axes, draw the graphs of
  i $y = f(x) - 3$ ii $y = f(-x)$ iii $y = f(x + 1)$

b Describe fully each transformed graph.

8 Work out the algebraic equation of each transformed graph in **Q7**. Simplify fully.

9 Reasoning

a Make three seperate sketches of the graph of $y = \cos x$ for $0° \leqslant x \leqslant 360°$

b On these sketches, sketch the graphs of

i $y = 2\cos x$ ii $y = \cos(x - 90°)$ iii $y = \cos(2x)$

> **Q9 hint** $2\cos x = 2f(x)$
> $\cos(x - 90°) = f(x - 90°)$
> $\cos(2x) = f(2x)$
> Apply each transformation.

10 Reasoning / Problem-solving This is the graph of $y = f(x) = x^3 - 3x + 2$

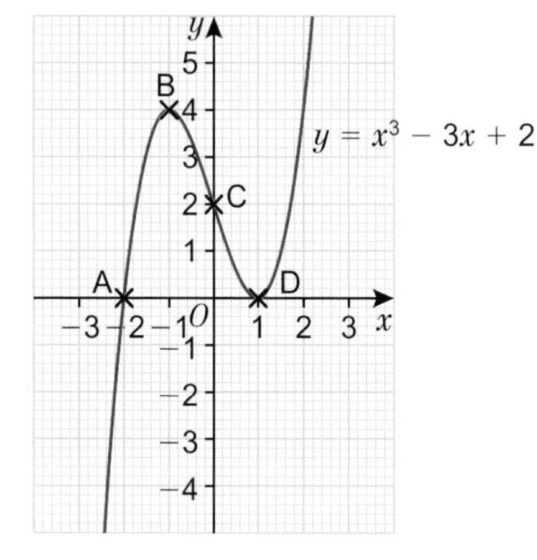

a Points, A(−2, 0), B(−1, 4), C(0, 2) and D(1, 0) lie on this cubic graph.
Work out the positions of points A, B, C and D after each of these transformations.

i $y = f(x) - 4$ ii $y = f(x - 4)$ iii $y = f(-x)$ iv $f = -f(x)$

v $y = 2f(x)$ vi $y = f(2x)$ vii $y = f(x - 1) - 3$ viii $y = f(x + 2) + 3$

b Sketch the transformed graphs in part **vii** $y = f(x - 1) - 3$ and part **viii** $y = f(x + 2) + 3$

c Describe fully the transformations in part **vii** $y = f(x - 1) - 3$ and part **viii** $y = f(x + 2) + 3$

2.8 Simultaneous equations

Objective

- Solve simultaneously one linear and one quadratic equation.

Key point 29

A linear graph and a quadratic graph can intersect at 0, 1 or 2 points.

- Line 1 intersects the quadratic graph at points A and B.
- Line 2 is a tangent to the curve and just touches it at point C.
- Line 3 has no points of intersection with the quadratic graph.

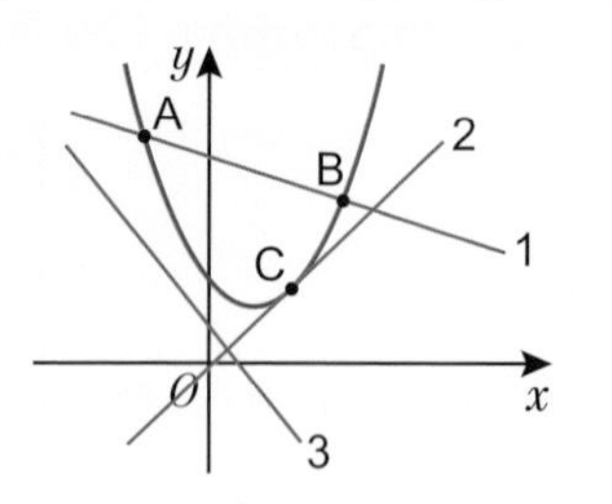

Key point 30

Solving a linear equation and a quadratic equation simultaneously by an algebraic method will give exact values for the coordinates of the points of intersection of the two graphs.

Key point 31

To solve a linear equation and a quadratic equation simultaneously, first rearrange the linear equation to make x or y the subject and substitute into the quadratic equation. Then solve the resulting quadratic equation by factorising, by using the quadratic formula or by completing the square.

1 Factorise

a $x^2 - 49$
b $2x^2 - 200$
c $x^2 - 7x - 18$
d $3x^2 - 17x + 20$

2 Make y the subject.

a $2y = 4x - 3$
b $x - 3y = 6$
c $3x - 5y = 20$
d $5 + 2y = 4x$

3 Expand and simplify.

a $(x - 4)^2$
b $(x + 5)^2$
c $(7 - x)^2$
d $(5x - 4)(3x + 8)$

4 Solve these simultaneous equations.

a $y = x + 12$
$y = x^2$

b $y = 5x$
$y = x^2 - 24$

c $y^2 = 6x$
$y = 2x - 6$

Q4a hint Simply equate, or substitute y from the linear equation into the quadratic one, giving $x^2 = x + 12$. Then solve for x and y.

Q4 A common error is to not realise that the values for x and y need to be correctly paired.

5 Solve these simultaneous equations.

a $y = x + 4$
$y = x^2 - 2x$

b $y = 2x - 3$
$y = 2x^2 - x - 23$

c $y = 5x + 2$
$y = x^2 + x - 10$

6 Solve these simultaneous equations.

a $y = 2x - 5$
$x^2 + y^2 = 25$

b $y = 3x + 2$
$y^2 = 2x^2 - 1$

c $xy = 14$
$4x - 3y = 17$

Q6c hint Write $x = \frac{14}{y}$ or $y = \frac{14}{x}$ and substitute into the other equation. Rearrange to give a quadratic equation in x or y.

7 Reasoning $y = x^2 - 2x + 5$ is a quadratic graph.
$y = 2x + 1$ is a straight-line graph.
Do these graphs intersect at 0, 1 or 2 points?
Show working to justify your answer.

8 Reasoning A circle has equation $x^2 + y^2 = 16$
The straight line $y = x + 3$ intersects the circle in two places.
Work out the coordinates of the points of intersection.
Give your answers correct to 2 d.p.

9 Reasoning Do these simultaneous equations have 0, 1 or 2 solutions?
$2x + y = 10$
$y = 6 - 5x - x^2$
Show working to justify your answer.
Illustrate your answer by drawing a sketch graph.

10 Reasoning The sketch graph shows the curve $y = \frac{6}{x}$ and the straight line $y = 12 - 2x$

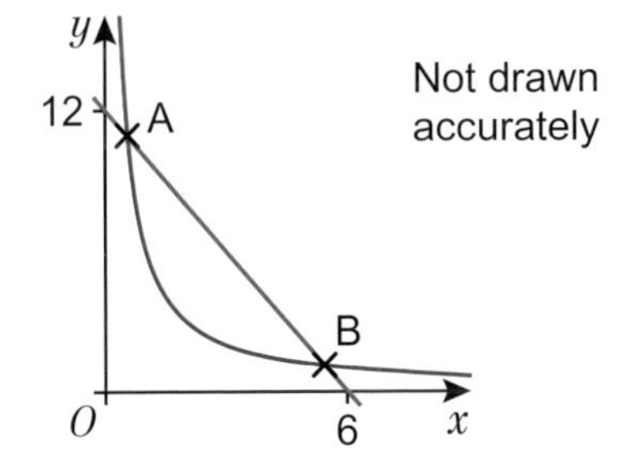

The graphs intersect at the points marked A and B.
Work out the coordinates of the points A and B. Give your answers in surd form.

Warm up

2.9 Equations, expressions and inequalities

Objectives

- Factorise quadratic expressions of the form $ax^2 + bx + c$.
- Expand products of three binomial expressions.
- Simplify algebraic expressions involving algebraic fractions.
- Solve equations involving algebraic fractions.
- Solve quadratic inequalities.
- Find approximate solutions to quadratic equations numerically, using iteration.

Key point 32

To find the product of three **binomial expressions**, multiply two of them to give a quadratic expression, then multiply each term of this quadratic by the terms in the third binomial. Simplify fully to give a cubic expression.

Key point 33

You may need to factorise before simplifying an **algebraic fraction**.

- Factorise the numerator and denominator.
- Divide the numerator and denominator by any common factors.

Key point 34

You may need to factorise the numerator and/or denominator before you multiply or divide algebraic fractions.

Key point 35

You can add or subtract algebraic fractions by first factorising the denominators then rewriting the fractions using a common denominator. Fully simplify your final answer.

Key point 36

When an equation involves algebraic fractions, first identify a common denominator, then multiply *all* terms of the equation by the common denominator. Then expand any brackets, simplify and rearrange into a quadratic equation of the form $ax^2 + bx + c = 0$.

Key point 37

To solve a **quadratic inequality**, first factorise, then sketch a graph of the function, and then use **set notation** to identify and describe the required regions.

Key point 38

To find an accurate root of a quadratic equation you can use an **iterative process**.
Iterative means carrying out a repeated action.
Follow these steps.

1. Rearrange the quadratic equation to make x^2 the subject, then take the square root of both sides. This gives you your iterative formula.
2. Sustitute your initial value (x_0) in the expression with the square root and work out your next value (x_1).
3. Without rounding any of your answers, repeat this process until two consecutive answers are the same, when rounded to the required degree of accuracy.

Warm up

1 Expand

a $(x+4)(x-9)$ b $(2x-7)^2$ c $(3x+2y)^2$

2 Factorise

a $5x-35$ b $4x^2-25y^2$ c $x^2-17x+60$ d $2x^2-3x-44$

3 Simplify

a $\frac{2x}{3}+\frac{x}{4}$ b $\frac{2}{a}+\frac{5}{2a}$

4 Expand

a $(x+5)(x+3)$ b $(x+5)(x+3)(x-4)$ c $(x-3)(x+7)(x+2)$

5 Expand

a $(x-3)(x+2)(x+6)$ b $(x-4)(x+1)(x-8)$
c $(x-5)(x-4)(3x+2)$ d $(3x-4)(x+2)(2x-5)$

Q5 hint You can multiply any two brackets first – you do not have to do them in the order in which they are written.

6 Simplify

a $\frac{x+4}{5x+4}$ b $\frac{(x+2)(x+6)}{(x-1)(x+2)}$
c $\frac{x^2(x+3)}{x(x+3)^2}$

Q6 hint Do not expand the brackets – cancel common factors first. Remember, you can only cancel whole brackets.

7 Simplify fully

a $\frac{x^3+6x^2}{x}$ b $\frac{12x^2-8x}{3x-2}$ c $\frac{3(x-4)}{x^2+2x-24}$
d $\frac{x^2-8x+15}{x^2+7x-30}$ e $\frac{x^2-64}{x^2+9x+8}$ f $\frac{x^2-3x-54}{x^2-81}$

8 Simplify fully

a $\frac{2x^2-x-10}{3x^2+5x-2}$ b $\frac{4x^2+16x+7}{4x^2+4x-35}$ c $\frac{6x^2-7x-24}{2x^2-5x-12}$

Q8 A common error is to make mistakes in cancelling after factorising.

9 Write each of these as a single fraction in its simplest form.

a $\frac{w^2-9}{3w+12}\times\frac{2w+8}{w^2-w-6}$ b $\frac{m^2-4}{4}\times\frac{8}{m^2+5m+6}$
c $\frac{x+1}{x^2+7x+12}\div\frac{x^2+2x+1}{x+4}$ d $\frac{9t^2-4}{t+2}\div\frac{3t^2+5t+2}{t^2+3t+2}$

Q9c hint Invert the second fraction and multiply.
$\frac{a}{b}\div\frac{c}{d}=\frac{a}{b}\times\frac{d}{c}$

10 Write each of these as a single fraction.

a $\frac{x+1}{3}+\frac{x-6}{2}$ b $\frac{1}{xy}+\frac{3}{2y}$
c $\frac{4}{x^2}-\frac{2}{xy}$ d $\frac{3x}{4y}-\frac{y}{2x}$

Q10 hint First, find the LCM of the terms in the denominator. Always fully simplify your answer.

11 Write each of these as a single fraction.

a $\frac{3}{x+2}+\frac{1}{x+4}$ b $\frac{6}{x+3}+\frac{5}{x-2}$ c $\frac{4}{x+1}-\frac{3}{x-5}$ d $\frac{2}{x-3}-\frac{5}{x-1}$

12 Write each of these as a single fraction.

a $\frac{3}{2x^2-3x-9}+\frac{4}{x^2-9}$ b $\frac{4}{x^2+4x-21}-\frac{2}{x^2+11x+28}$ c $\frac{5}{3x^2-10x+8}-\frac{1}{3x^2-x-4}$

13 Solve

a $\frac{5}{x} + \frac{2}{x+2} = 3$ b $\frac{3}{x} - \frac{3}{x+3} = \frac{9}{10}$ c $\frac{1}{x+3} + \frac{3}{3x+1} = 1$ d $\frac{7}{x-3} - \frac{6}{x-1} = 2$

14 Solve

a $\frac{9}{x} - \frac{10}{x+2} = 7$ b $\frac{1}{2x-1} + \frac{x+1}{3x} = \frac{5}{6}$ c $\frac{4}{3x-1} + \frac{5}{x-1} = 3$ d $\frac{5}{x-4} - \frac{10}{2x-5} = 3$

15 Solve, giving your answers correct to 2 d.p.

a $\frac{7}{x+3} + \frac{1}{x-2} = 2$ b $\frac{5}{x-4} - \frac{3}{2x-3} = 1$

16 **Reasoning** For what values of x are these inequalities satisfied? Give your answers using set notation.

a $x^2 - 11x - 60 \leqslant 0$ b $2x^2 - 5x + 2 > 0$

c $5 + x - 4x^2 > 0$ d $19x \leqslant 3x^2 + 20$

Q16 hint Use the symbol ∪ to show that the solution includes all the values satisfied by two different inequalities.

17 **Reasoning** For what values of x are these inequalities satisfied? Give your answers using set notation.

a $12x^2 > x + 6$ b $3 - x \geqslant 10x^2$ c $9x^2 + 10 < 21x$ d $28 \leqslant 20x^2 + 19x$

18 a Show that the equation $x^2 - 3x - 5 = 0$ can be rewritten in the form $x = \sqrt{3x + 5}$

b Using $x = \sqrt{3x + 5}$ with $x_0 = 4$, use iteration to find one root of the equation $x^2 - 3x - 5 = 0$, giving your answer correct to 5 d.p.

Q18b hint Look at the steps in Key point 38.

19 Use iteration to find one root of each of these equations, giving your answers correct to 5 d.p.

a $x^2 - 4x - 4 = 0$, use $x_0 = 5$ b $x^2 - 6x - 8 = 0$, use $x_0 = 7$

20 **Reasoning** Tony says that the number 351 is divisible by 9 because the sum of the digits $(3 + 5 + 1)$ is 9.

a Verify that 351 is divisible by 9.

b Prove that Tony's result is always true.

Q20b hint You can write a 3-digit number 'xyz' as $100x + 10y + z$.

2.10 Sequences

Objectives

- Recognise and use Fibonacci-type sequences, geometric sequences and quadratic sequences.
- Deduce an expression for the nth term of a quadratic sequence.

Key point 39

In a **Fibonacci-type sequence**, the next number is found by adding the previous two numbers together. For example, 1, 1, 2, 3, 5, 8, 13, 21, … is a Fibonacci-type sequence because $1 + 1 = 2$, $1 + 2 = 3$, $2 + 3 = 5$ and so on.

Key point 40

In a **geometric sequence**, the terms increase or decrease by a **constant multiplier**, known as the **common ratio** (r). The nth term of a geometric sequence with first term a is ar^{n-1}.

Key point 41

A **quadratic sequence** has n^2 and no higher power of n in the nth term.

Key point 42

The **second differences** of a quadratic sequence, $u_n = an^2 + bn + c$, are constant and are equal to $2a$.

Key point 43

The ***n*th term** of a quadratic sequence can be worked out in three steps.

1. Work out the second differences.
2. Halve the second difference to find a, the coefficient of n^2.
3. Subtract the sequence an^2. You may need to add a constant, or find the nth term of the remaining terms.

1 Write down the next three terms in each of these Fibonacci-type sequences.

a 1, 5, 6, 11, □, □, □ b 1, 3, 4, 7, □, □, □ c 2, 5, 7, 12, □, □, □

2 Write down the next three terms in each of these geometric sequences.

a 1, 4, 16, 64, □, □, □ b −2, −6, −18, −54, □, □, □ c 160, −80, 40, −20, □, □, □

3 A quadratic sequence has nth term $u_n = 2n^2 - 11n + 7$.
Write down the first five terms of the sequence.

4 **Reasoning** Work out the missing numbers in each of these Fibonacci-type sequences.

a 3, □, □, □, □, 54
b 4, □, □, □, □, □, 124
c 5, □, □, □, □, □, □, −12
d 11, □, □, □, □, □, □, −16

Q4a hint Let the second term be $3 + x$, then the third term will be $3 + 3 + x = 6 + x$. Use algebra to work out the other terms and set up an equation.

5 **Problem-solving** A Fibonacci-type sequence has seven terms.
The first term is n and the 7th term is n.
Work out all the terms of the sequence in terms of n.

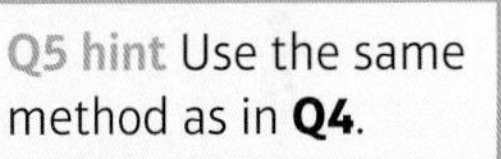

Q5 hint Use the same method as in **Q4**.

6 Work out the common ratio and the 8th term in each of these geometric sequences.

a 8, 12, 18, 27, ...
b 243, −81, 27, −9, ...
c 3, $\sqrt{18}$, 12, $\sqrt{288}$, ...

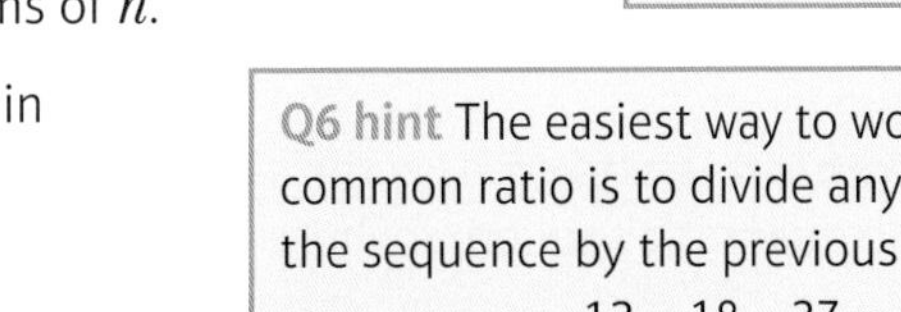

Q6 hint The easiest way to work out the common ratio is to divide any term of the sequence by the previous term.
For part **a**, $r = \frac{12}{8} = \frac{18}{12} = \frac{27}{18} = \square$

7 The number of fish in a lake is decreasing by 30% each year because of pollution.
There are about 50 000 fish now.
Predict the number of fish in 10 years' time.

8 **Reasoning** A geometric sequence has a 4th term of $6\sqrt{3}$ and a 5th term of 18.

a Work out the common ratio.
b Deduce the first term of the sequence.
c Work out the 10th term of the sequence.

9 **Reasoning** The first three terms of a geometric sequence are $K + 6$, K and $K - 2$.

a Explain clearly why $\frac{K}{K+6} = \frac{K-2}{K}$
b Work out the value of K.
c Deduce the 8th term of the geometric sequence, giving your answer as a fraction.

10 Copy and complete these diagrams to work out first and second differences for each of these quadratic sequences.

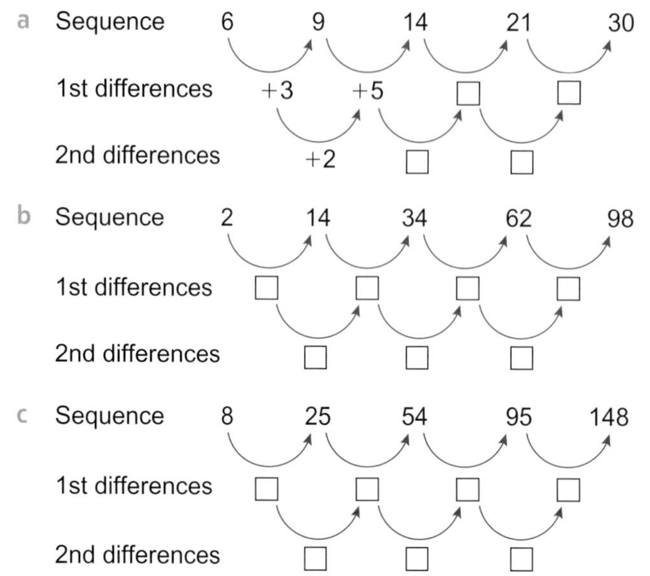

11 Reasoning Work out a formula for the nth term of each quadratic sequence.

a 5, 11, 21, 35, 53, ...

b 8, 17, 32, 53, 80, ...

c 6.5, 8, 10.5, 14, 18.5, ...

> **Q11 hint** Follow the steps in Key point 43. Remember to halve the second difference to get the coefficient a in the expression $an^2 + bn + c$.

12 Problem-solving The nth term of a quadratic sequence is $u_n = an^2 + bn + c$
The first three terms of the sequence are 6, 7 and 12.

a Explain why $a + b + c = 6$ (let this be equation ①).

b Write down two more equations in a, b and c (let these be equations ② and ③).

c Subtract equation ① from equation ② and subtract equation ② from equation ③.

d Solve the resulting simultaneous equations to find the values of a and b.

e Deduce the value of c.

f State the nth term of the quadratic sequence.

13 Problem-solving Use the same method as in **Q12** to work out a formula for the nth term of the quadratic sequence whose first three terms are 2, 6 and 20.

2.11 Mixed exercise

Objective

- Consolidate your learning with more practice.

1 Problem-solving Work out the equation of the line that is perpendicular to $5x - 10y + 4 = 0$ and passes through $(-3, 15)$.

2 a Expand $(6m - 5t)(2m + 3t)$

b Factorise $75x^2 - 12y^2$

3 Reasoning

a Sketch the graphs of $y = \frac{12}{x}$ and $2y - x = 2$

b Work out the coordinates of the points where these graphs intersect.

4

Exam question

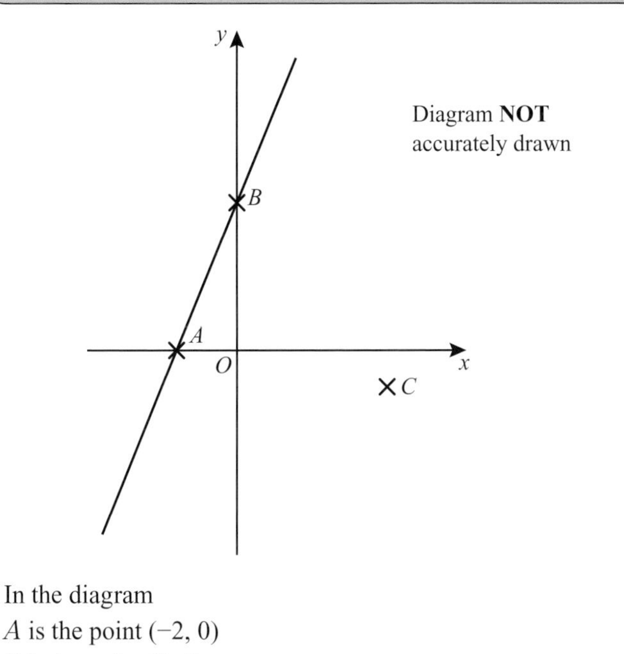

In the diagram
A is the point $(-2, 0)$
B is the point $(0, 4)$
C is the point $(5, -1)$
Find an equation of the line that passes through C and is perpendicular to AB. **(4 marks)**

November 2012, Q16, 5MB2H/01

Exam hint
Find the gradient m of the line AB. The line perpendicular to AB has gradient $-\frac{1}{m}$

Q4 A common error is to find the gradient of the perpendicular line but then to fail to find c and so work out the equation of the line. Remember to read the question carefully.

5

Exam question

a Complete the table of values for $y = x^2 - 2x - 1$

x	−2	−1	0	1	2	3	4
y	7			−2	−1		

b On the grid, draw the graph of $y = x^2 - 2x - 1$ for values of x from −2 to 4.

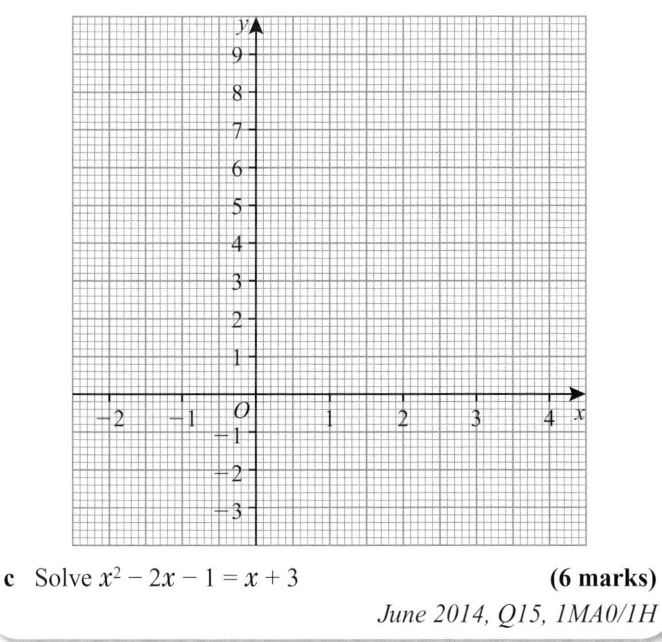

c Solve $x^2 - 2x - 1 = x + 3$ **(6 marks)**

June 2014, Q15, 1MA0/1H

Exam hint
Your graph should have at least 6 points plotted correctly.

6 **Exam question**

Here are three graphs.

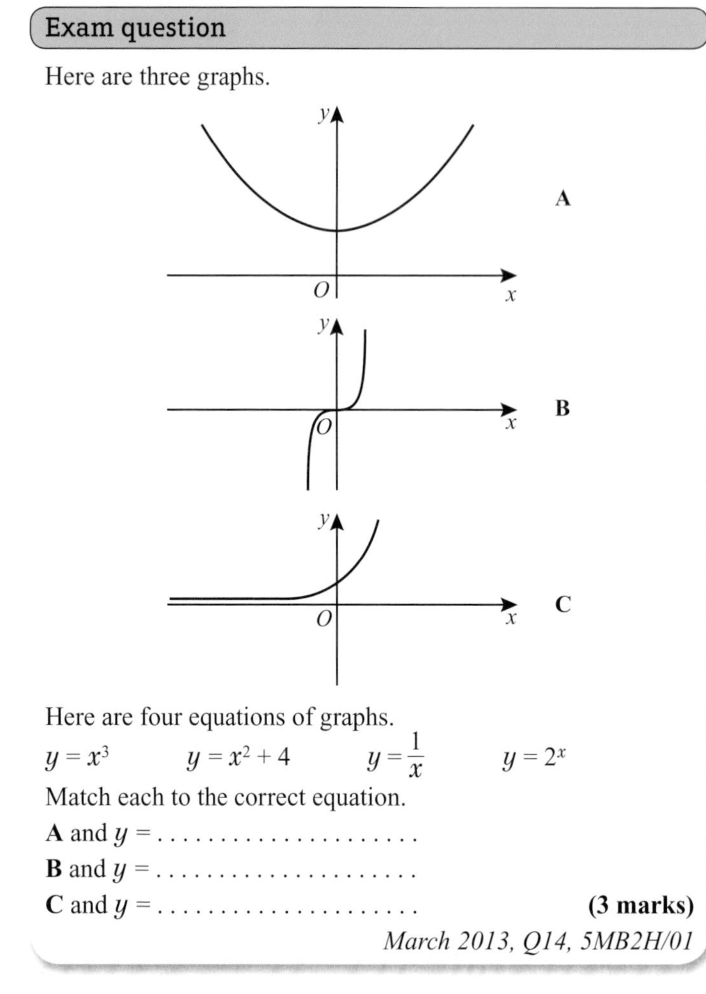

Here are four equations of graphs.

$y = x^3$ $\quad$ $y = x^2 + 4$ $\quad$ $y = \frac{1}{x}$ $\quad$ $y = 2^x$

Match each to the correct equation.

A and y = ...

B and y = ...

C and y = ... **(3 marks)**

March 2013, Q14, 5MB2H/01

7 **Problem-solving**

a Copy and complete the table of values for $y = 3x^2 - 4x - 2$

x	−2	−1	0	1	2	3
y		5				13

b Draw the graph of $y = 3x^2 - 4x - 2$ for $-2 \leqslant x \leqslant 3$.

c Use your graph to solve the equation $3x^2 - 4x - 2 = 0$

d Solve the equation $3x^2 - 2x - 7 = 0$

8 For this question, use the quadratic graph you drew in **Q7**.

a Work out the gradient of the graph when $x = -0.5$.

b Work out the gradient of the graph when $x = 2$.

9 **Reasoning**

a Copy and complete the table of values for $y = 16 - x^2$.

x	−4	−3	−2	−1	0	1	2	3	4
y		7			16			7	0

b Draw the graph of $y = 16 - x^2$ for $-4 \leqslant x \leqslant 4$.

c Estimate the area under the graph and bounded by the x-axis, from $x = -4$ to $x = 4$.

10 **Problem-solving** A circle has equation $x^2 + y^2 = 52$

a Write down the radius of the circle.

b Work out the equation of the tangent to the circle at the point (−6, 4).

11 Reasoning You are given that $\cos 72° = 0.3090$

a Write down another angle in the range $0° \leqslant x \leqslant 360°$ for which $\cos x = 0.3090$

b Solve $\cos x = -0.3090$ for $0° \leqslant x \leqslant 360°$.

12 Solve these simultaneous equations.

a $y = 2x + 9$
$y = 2x^2 - 15x$

b $y^2 = x$
$y = 1 - 2x$

13 Expand

a $(x - 6)(x - 5)(x + 4)$

b $(2x - 3)(x - 2)(x + 7)$

14 Solve

a $\frac{9}{2x - 7} + \frac{6}{x - 1} = 3$

b $\frac{11}{4x - 5} - \frac{8}{x + 1} = 1$

15 Use iteration to find one root of the quadratic equation $x^2 - 2x - 9 = 0$.
Use $x_0 = 4$
Give your answer correct to 5 d.p.

16 Reasoning Work out a formula for the nth term of the quadratic sequence whose first three terms are

a 8, 7 and 12

b −5, 9 and 31

17

Exam question

The expression $x^2 - 8x + 21$ can be written in the form $(x - a)^2 + b$ for all values of x.

a Find the value of a and the value of b.

a = ...

b = ...

The equation of a curve is $y = f(x)$ where $f(x) = x^2 - 8x + 21$
The diagram shows part of a sketch of the graph of $y = f(x)$.

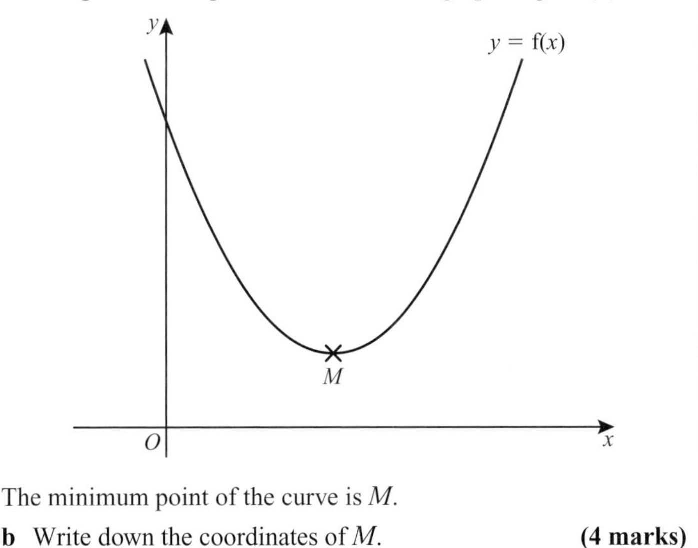

The minimum point of the curve is M.

b Write down the coordinates of M. **(4 marks)**

June 2013, Q25, 1MA0/1H

Exam hint
You must be able to recognise that this question requires you to complete the square.

Q17 A common error is to not use the values from part **a** to answer part **b**. When a quadratic is written in the form $a(x + b)^2 + c$, the turning point is $(-b, c)$.

18 **Exam question**

Solve the simultaneous equations

$x^2 + y^2 = 9$

$x + y = 2$

Give your answers correct to 2 decimal places.

x = ... y = ...

or x = ... y = ... **(6 marks)**

June 2013, Q25, 1MA0/1H

Exam hint
If the question asks you for an answer correct to 2 decimal places, you should recognise you will need to use the quadratic formula.

19 **Exam question**

Solve the equations

$x^2 + y^2 = 36$

$x = 2y + 6$ **(5 marks)**

June 2014, Q26, 1MA0/2H

Exam hint
Choose the easiest substitution and take care when expanding.

20 **Exam question**

a Solve $\frac{4(8x - 2)}{3x} = 10$

b Write as a single fraction in its simplest form

$\frac{2}{y+3} - \frac{1}{y-6}$ **(6 marks)**

November 2013, Q20, 1MA0/1H

Exam hint
Find the common denominator and express each fraction in terms of the common denominator before subtracting.

21 a Write $2x^2 - 8x + 13$ in the form $a(x + b)^2 + c$ where a, b and c are integers.

b Sketch a graph of the function $y = 2x^2 - 8x + 13$ and identify the turning point.

22 **Reasoning** $f(x) = 3x + 5$ $g(x) = 1 - 4x$ $h(x) = 2 - x^2$

Work out, simplifying fully

a $g^{-1}(x)$ b $gf(x)$ c $fh(x)$ d $hg(x)$

23 **Reasoning** Sketch these graphs, marking clearly the points of intersection with the x- and y-axes.

a $y = (x + 7)(x - 2)(x + 2)$ b $y = (x - 4)(x + 1)^2$ c $y = 6x - 5x^2 - x^3$

24 **Problem-solving** Solve the equation $2 \tan x + 9 = 0$ for $-180° \leqslant x \leqslant 180°$.
Give your answers correct to 1 d.p.

25 **Reasoning** The diagram shows a sketch graph of $y = f(x) = 7 + 6x - x^2$

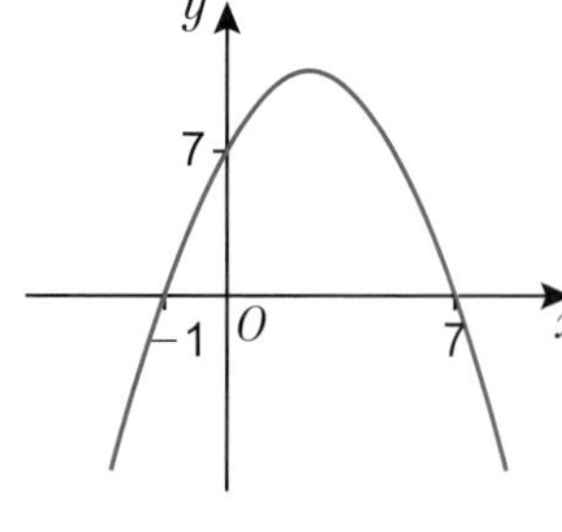

a Copy the diagram and, on the same axes, draw sketch graphs of

i $y = -f(x)$ ii $y = \frac{1}{2}f(x)$ iii $y = f(2x)$

b Describe fully each transformation.

c Work out the algebraic equation of each transformed graph.
Simplify fully.

26 Reasoning Do these simultaneous equations have 0, 1 or 2 distinct solutions? If there are 1 or 2 distinct solutions, solve them.
Give your answers correct to 2 d.p. where necessary.

a $y = 7x + 2$
$y = 5x^2 - 2x$

b $x^2 + y^2 = 10$
$y = 5 + x$

c $3x = 2y - 4$
$y = 2x^2 - 10$

27 Write each of these as a single fraction in its simplest form.

a $\dfrac{x^2 + 2x - 15}{x^2 - 4} \times \dfrac{2x + 4}{x^2 - 7x + 12}$

b $\dfrac{4m^2 - 4m - 3}{m^3 + 8m^2} \div \dfrac{6m^2 - 5m - 6}{m^2 + 8m}$

28 Problem-solving The diagram shows a sketch of $y = \sin(2x)$

Q28 hint Answers can be given in terms of degrees of in terms of multiples of π.

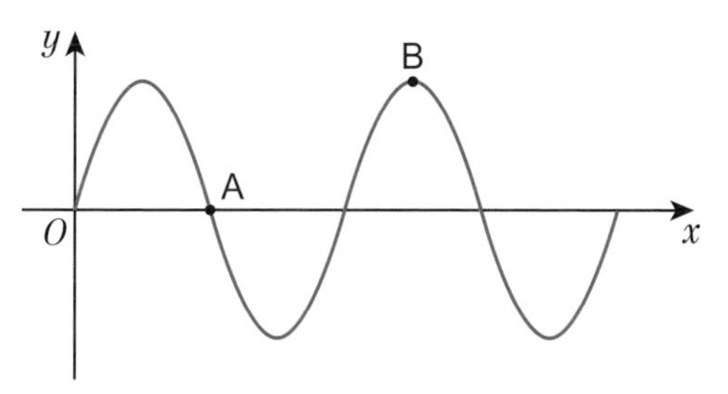

a Write down the coordinates of A.

b Write down the coordinates of B.

The diagram shows a sketch of $y = a + b\sin(cx)$

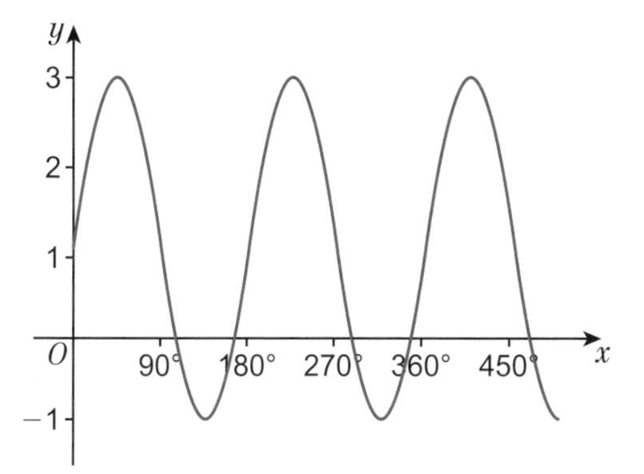

c Find the values of a, b and c.

29 Reasoning $f(x) = 5(x - 3)$ $g(x) = 5(x + 3)$

a Find $f^{-1}(x)$.

b Find $g^{-1}(x)$.

c Work out $f^{-1}(x) + g^{-1}(x)$.

d If $f^{-1}(a) + g^{-1}(a) = 4$ work out the value of a.

30 **Exam question**

The first three terms of a geometric series are $(k + 4)$, k and $(2k - 15)$ respectively, where k is a positive constant.

a Show that $k^2 - 7k - 60 = 0$. **(4 marks)**

b Hence show that $k = 12$. **(2 marks)**

c Find the common ratio of this series. **(2 marks)**

January 2009, Q9 (parts a–c), A level paper GCE Mathematics (6664/01)

Exam hint
'Show that …', means you must show every step of your working to show *how* to achieve the given result.

31 **Exam question**

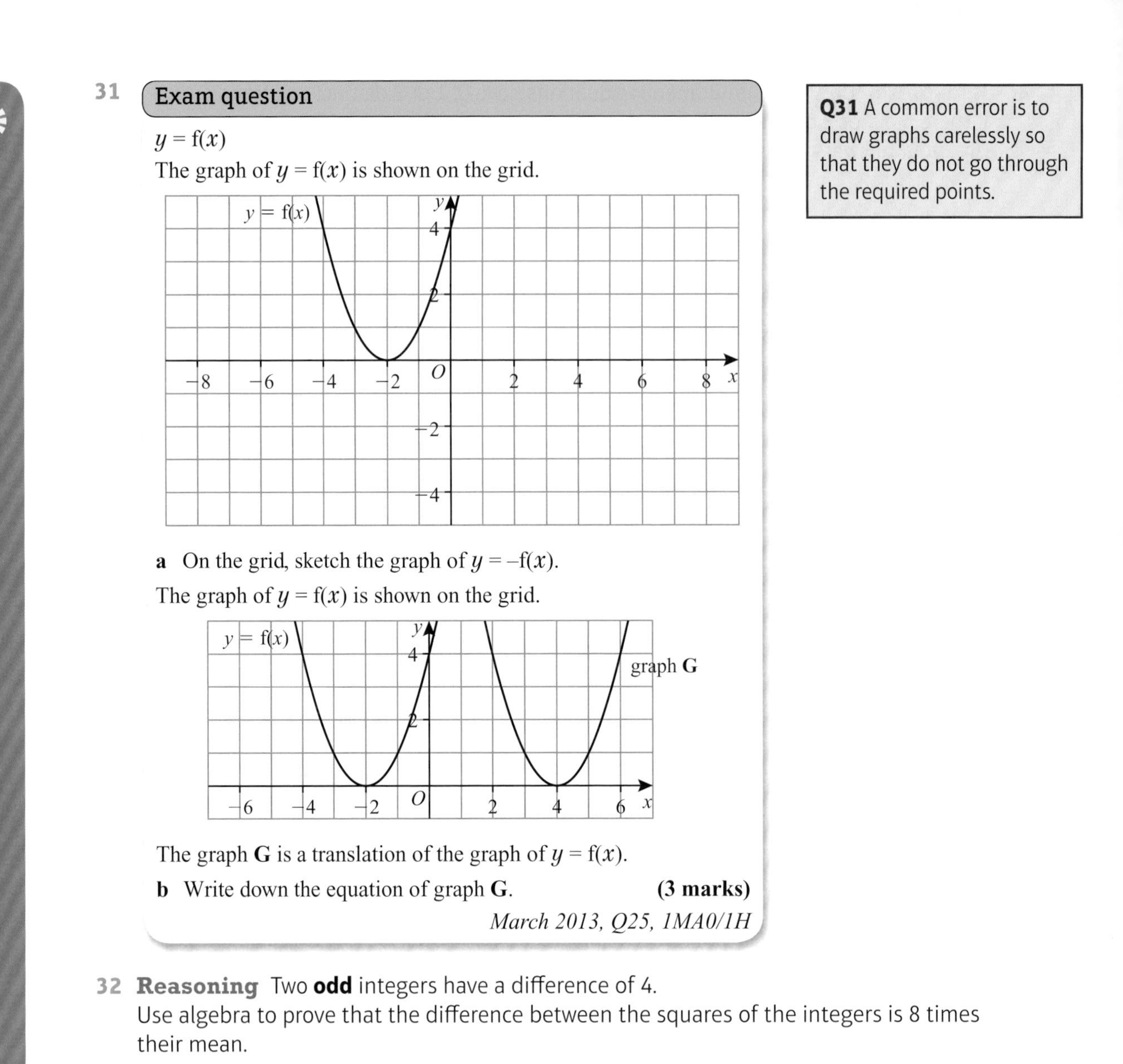

$y = f(x)$

The graph of $y = f(x)$ is shown on the grid.

a On the grid, sketch the graph of $y = -f(x)$.

The graph of $y = f(x)$ is shown on the grid.

The graph **G** is a translation of the graph of $y = f(x)$.

b Write down the equation of graph **G**. **(3 marks)**

March 2013, Q25, 1MA0/1H

Q31 A common error is to draw graphs carelessly so that they do not go through the required points.

32 **Reasoning** Two **odd** integers have a difference of 4.
Use algebra to prove that the difference between the squares of the integers is 8 times their mean.

33 **Reasoning** $f(n) = n^2 - 3n$
Prove that $f(2n) - f(n + 1) - f(n + 2) \equiv 2f(n) + 4$

3 RATIO, PROPORTION AND RATES OF CHANGE

What is covered in this section	Higher Student Book reference
3.1 Compound measures • Solve problems involving compound measures.	Unit 11
3.2 Area and volume scale factors • Use the links between length, area and volume scale factors to solve problems.	Units 8, 12
3.3 Proportion • Write and use equations to solve problems involving direct and inverse proportion.	Units 11, 19
3.4 Exponential graphs • Recognise, sketch and interpret graphs of exponential functions.	Unit 19
3.5 Mixed exercise • Consolidate your learning with more practice.	

3.1 Compound measures

Objective

- Solve problems involving compound measures.

Key point 1

Compound measures combine measures of two different quantities.

Speed is a measure of **distance** travelled and **time** taken. It can be measured in metres per second (m/s), kilometres per hour (km/h) or miles per hour (mph).

Average speed = $\frac{\text{distance}}{\text{time}}$ or $S = \frac{D}{T}$

Key point 2

Density is the **mass** of substance in g contained in a certain **volume** in cm^3 and is often measured in grams per cubic centimetre (g/cm^3).

Density = $\frac{\text{mass}}{\text{volume}}$ or $D = \frac{M}{V}$

Key point 3

Pressure is the **force** in newtons applied over an **area** in cm^2 or m^2. It is usually measured in newtons (N) per square metre (N/m^2) or per square centimetre (N/cm^2).

Pressure = $\frac{\text{force}}{\text{area}}$ or $P = \frac{F}{A}$

Warm up

1 A 24 cm^3 block of silver has a mass of 252 g.

a Work out the density of silver.

b Work out the volume of a 483 g block of silver.

2 A bottle of water has a flat circular base of diameter 6 cm.
The bottle exerts a force of 5 N on the table.
Work out the pressure, in N/cm², of the bottle on the table.
Give your answer correct to 3 s.f.

Q2 hint Work out the area in contact with the table.

3 Petrol has a density of 0.71 g/cm³.
The fuel tank of a car holds 55 litres.
What is the mass of petrol in a full tank?
Give your answer in kg.

Q3 hint Be careful with the units.

4 A car travels at an average speed of 18 m/s for 7.4 seconds, both figures correct to 2 s.f.
Calculate the largest and the smallest possible distances travelled.

Q4 hint Distance = average speed × time.
Largest possible distance travelled = largest possible average speed × largest possible time.

5 A glass paperweight has a mass of 1360 g, correct to the nearest 10 g.
The volume of the paperweight is 390 cm³, correct to 2 s.f.
Work out the largest and smallest possible values for the density of the glass.
Give your answers correct to 3 s.f.

6 An iceberg of volume 17 000 m³ has a density of 0.93 g/cm³, both values correct to 2 s.f.
Work out the largest and smallest values of the mass of the iceberg, in tonnes.

Q6 hint Take care with units.

7 **Reasoning** A beaker has a capacity of 800 ml.
It contains 290 ml of oil.
An iron bar of mass 3.8 kg is placed in the beaker.
The density of iron is 7.9 g/cm³.
All measurements are correct to 2 s.f.
Show that there is a possibility of the beaker overflowing.

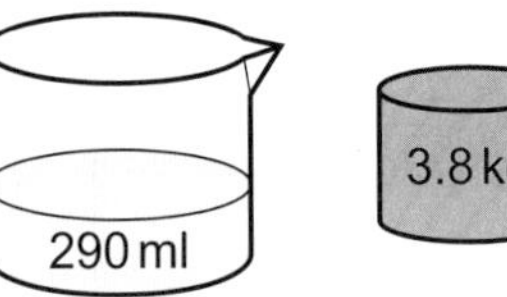

3.2 Area and volume scale factors

Objective

- Use the links between length, area and volume scale factors to solve problems.

Key point 4

When a shape is enlarged by a linear **scale factor** k, any areas are enlarged by scale factor k^2 and volumes are enlarged by scale factor k^3.

Key point 5

Shapes are **similar** when one shape is an enlargement of the other. Corresponding angles are equal and corresponding sides are all in the same ratio.

Warm up

1 Rectangle B is an enlargement of rectangle A by scale factor 3.
Rectangle A has a length of 8 cm.
Rectangle B has a width of 10.5 cm.
a Work out the perimeter of rectangle B.
b Work out the area of rectangle A.

Q1 hint Use the scale factor to work out the missing dimensions.

2 Here is a triangular prism.

a Work out the volume of an enlargement of this prism by scale factor 2.

b Work out the area of the sloping face (shaded) of the enlarged prism.

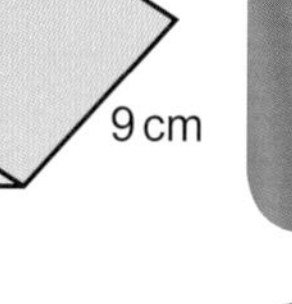

Q2 hint You will need to calculate the hypotenuse of the right-angled triangle.

3 These two shapes are mathematically similar.
The area of the smaller shape is 24 cm^2.
Work out the area of the larger shape.

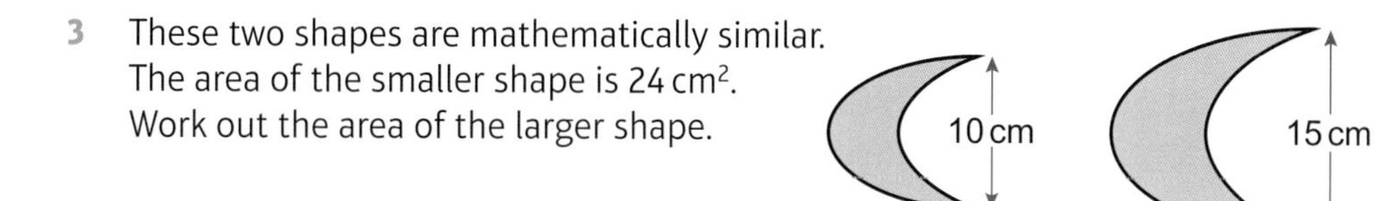

4 **Problem-solving** Two square-based pyramids A and B are mathematically similar.
The length of the base of pyramid A is 3.4 cm.
The length of the base of pyramid B is 8.5 cm.

a Work out the ratio of the surface areas of the two pyramids.

b The volume of the larger pyramid is 125 cm^3.
Work out the volume of the smaller pyramid.

Q4 hint First work out the linear scale factor.

5 **Problem-solving** On a map of scale 1 : 50 000 Grizedale forest has an area of 98 cm^2.
Work out the actual area of the forest.
Give your answer in km^2.

6 **Problem-solving** These prisms are mathematically similar.
The volume of prism A is 54 cm^3.
The volume of prism B is 1458 cm^3.
The area of the shaded face of the larger prism is 63 cm^2.

a Work out the length marked x cm.

b Work out the area of the shaded face of the smaller prism.

7 **Problem-solving** Two buckets are mathematically similar.
The diameter of the top of the smaller bucket is 14 cm.
The diameter of the top of the larger bucket is 31 cm.
The larger bucket is completely filled with water.
Assuming no spillage, how many of the smaller buckets can be filled from the larger bucket?
Show working to justify your answer.

8 **Problem-solving** Two perfume bottles are mathematically similar.
One holds 50 ml of perfume, the other holds 80 ml.
Work out the ratio of the surface areas of the two bottles.
Give your answer correct to 2 s.f.

9 **Problem-solving** Two prisms are mathematically similar.
The small prism has a length of 4.5 cm.
The large prism has a length of 10.8 cm.
The cross-sectional area of the large prism is 37.44 cm^2.

a Work out the cross-sectional area of the small prism.

b The two prisms are made out of the same material.
The small prism has a mass of 50 g.
Work out the mass of the large prism.

10 Problem-solving Two hot-air balloons are mathematically similar.
The smaller balloon has a volume of 1800 m^3. The larger balloon has a volume of 4500 m^3.
The area of fabric in the smaller balloon is 1500 m^2.
Work out the approximate area of the fabric in the larger balloon.

3.3 Proportion

Objective

- Write and use equations to solve problems involving direct and inverse proportion.

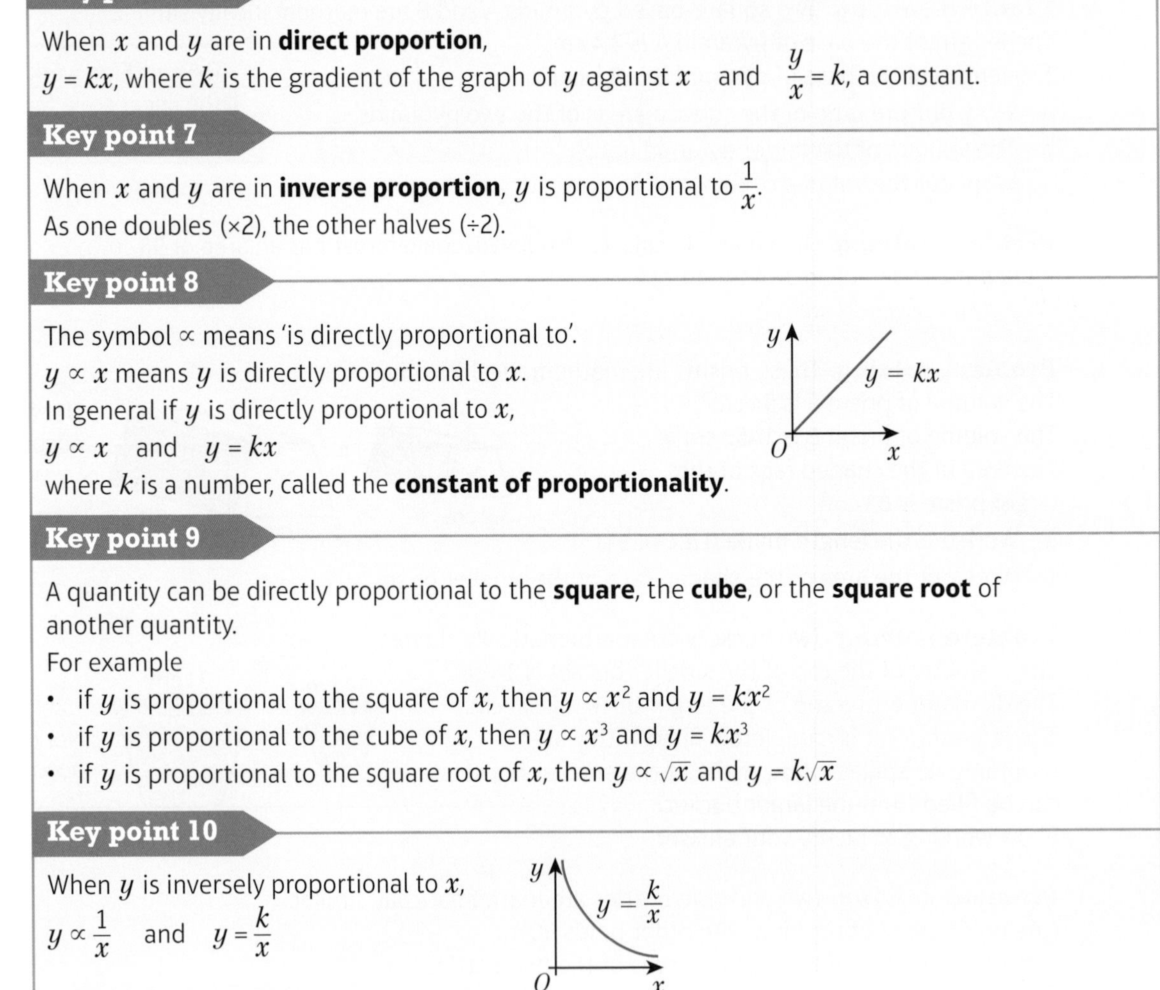

Key point 6

When x and y are in **direct proportion**,
$y = kx$, where k is the gradient of the graph of y against x and $\frac{y}{x} = k$, a constant.

Key point 7

When x and y are in **inverse proportion**, y is proportional to $\frac{1}{x}$.
As one doubles (×2), the other halves (÷2).

Key point 8

The symbol $\propto$ means 'is directly proportional to'.
$y \propto x$ means y is directly proportional to x.
In general if y is directly proportional to x,
$y \propto x$ and $y = kx$
where k is a number, called the **constant of proportionality**.

Key point 9

A quantity can be directly proportional to the **square**, the **cube**, or the **square root** of another quantity.
For example

- if y is proportional to the square of x, then $y \propto x^2$ and $y = kx^2$
- if y is proportional to the cube of x, then $y \propto x^3$ and $y = kx^3$
- if y is proportional to the square root of x, then $y \propto \sqrt{x}$ and $y = k\sqrt{x}$

Key point 10

When y is inversely proportional to x,
$y \propto \frac{1}{x}$ and $y = \frac{k}{x}$

Warm up

1 Mike and Maggie are going on holiday.
Mike changed £450 into euros and got €625.50.
Maggie changed £520 into euros and got €722.80.

a Work out the exchange rate for both Mike and Maggie.
Are these results consistent with pounds and euros being in direct proportion?

b Write a formula connecting pounds and euros.

c When they return they have €270.51 left altogether.
They convert this back to pounds at the rate of £1 = €1.42.
How much do they receive, in pounds?

2 For a constant force, F (N), the pressure, P (N/m²), is inversely proportional to the area, A (m²), that the force acts upon. When the area is 1.5 m², the pressure is 10.8 N/m². Work out the pressure exerted by this constant force when the area it acts on is a square of side 45 cm.

> **Q2 hint** Set up an equation connecting P and A using a constant of proportionality, k.

3 y is directly proportional to x.
$y = 21$ when $x = 6$

a Express y in terms of x.
b Work out y when $x = 11$
c Work out x when $y = 44.8$
d A graph of y against x will be a straight line.
What can you say about the gradient and y-intercept of this straight line?

> **Q3 hint** Set up an equation connecting y and x using a constant of proportionality, k, then work out k using the given values. Use your equation in all further calculations.

4 t is directly proportional to the square root of m.
$t = 37$ when $m = 9$

a Express t in terms of m. b Work out t when $m = 144$ c Work out m when $t = 18.5$

5 y is inversely proportional to x.
$y = 125$ when $x = 0.5$

a Express y in terms of x. b Work out y when $x = 0.2$ c Work out x when $y = 200$

6 w is inversely proportional to the square of p.
Copy and complete this table for values of p and w.

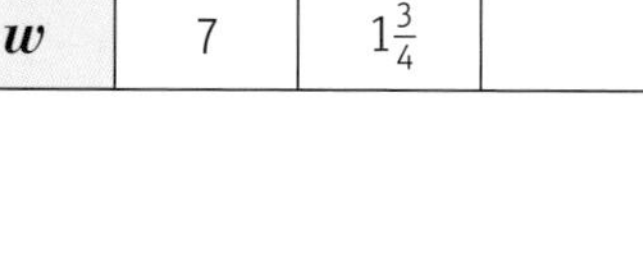

p	2		6
w	7	$1\frac{3}{4}$	

7 h is directly proportional to v^3.
$h = 30$ when $v = 2$

a Express h in terms of v. b Work out h when $v = \frac{4}{3}$ c Work out v when $h = \frac{10}{9}$

8 This triangular prism has a volume of 640 cm³.

a Work out a formula for y in terms of x.
b Describe the proportionality relationship between x and y.
c Work out y when $x = 4$ cm.
d Work out x when $y = 20$ cm.

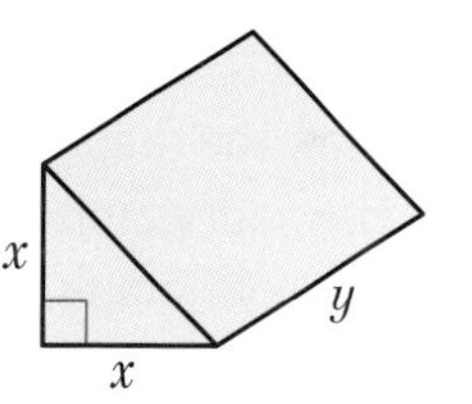

9 **Reasoning** Here are four graphs of y against x.

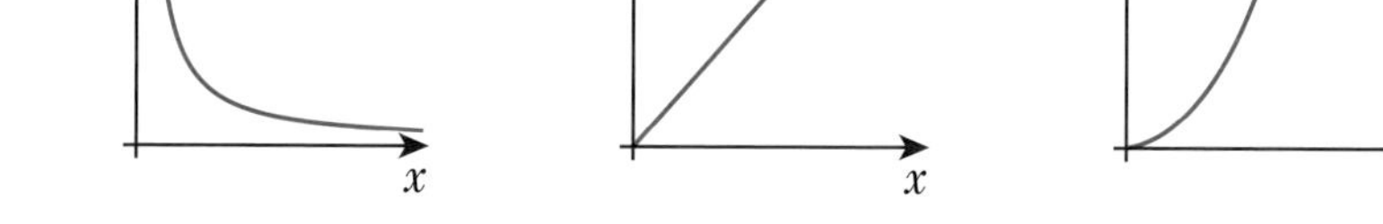

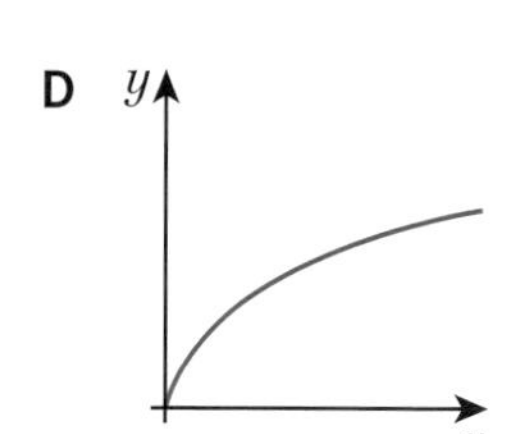

Match each of these proportionality relationships to the correct graph.

a $y \propto x$ b $y \propto \sqrt{x}$ c $y \propto \frac{1}{x}$ d $y \propto x^2$

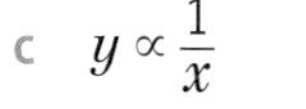

10 **Reasoning** In an experiment, measurements of t and w are taken. The table shows some results.
The variables, t and w,are connected by one of these rules.

t	0.9	5	12
w	400	12.96	2.25

A $w \propto \frac{1}{t}$ **B** $w \propto \frac{1}{t^2}$ **C** $w \propto \frac{1}{\sqrt{t}}$

Which is the correct rule?
Show working to justify your answer.

3.4 Exponential graphs

Objective

- Recognise, sketch and interpret graphs of exponential functions.

Key point 11

Expressions of the form a^x, where a is a positive number, are called **exponential functions**.

Key point 12

The graph of an exponential function has one of these shapes.

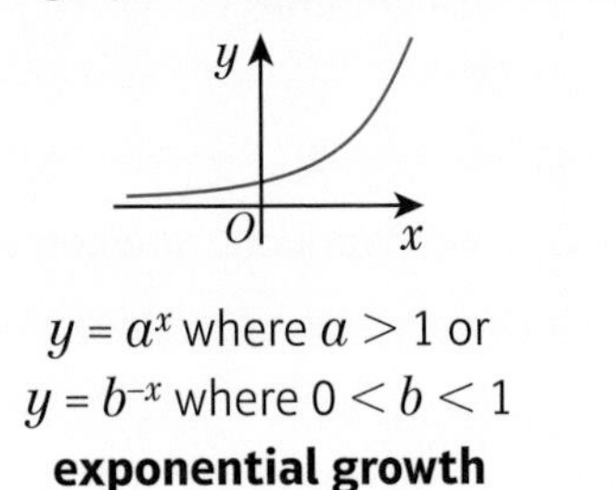

$y = a^x$ where $a > 1$ or
$y = b^{-x}$ where $0 < b < 1$
exponential growth

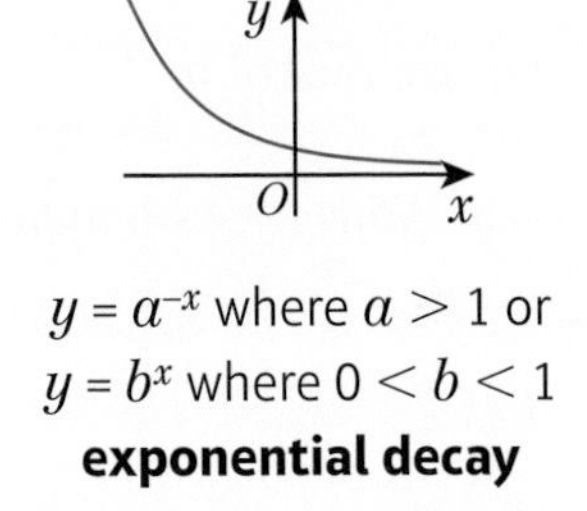

$y = a^{-x}$ where $a > 1$ or
$y = b^x$ where $0 < b < 1$
exponential decay

Warm up

1 Write down the value of x satisfying each of these equations.
a $2^x = 128$
b $3^x = 1$
c $4^x = \frac{1}{16}$
d $\left(\frac{1}{5}\right)^x = 625$

2 a Copy and complete the table of values for $y = \left(\frac{1}{2}\right)^x$

x	−3	−2	−1	0	1	2	3
y	8				0.5		0.125

b Draw the graph of $y = \left(\frac{1}{2}\right)^x$ for $-3 \leqslant x \leqslant 3$
c Use your graph to estimate the value of y when $x = 1.5$
d Use your graph to estimate the value of x when $y = 6$

Q2c, d hint Choose scales for both axes that enable you to read off the answers as accurately as possible.

3 **Reasoning** Katie invests £1500 in an account that pays 6% compound interest per year. The interest is added at the end of each year.
Assume that no money is added or withdrawn.
a Show that the total amount (T) after n years is given by the formula $T = 1500 \times 1.06^n$
b Draw the graph of $T = 1500 \times 1.06^n$ for $0 \leqslant n \leqslant 10$
c How many years does it take for the amount to exceed £2500?

Q3c hint You can use your graph or try different values of n in the formula. Give your answer as a whole number of years.

4 **Reasoning** A sample of radioactive material decays according to the formula $N = 70 \times 2^{\frac{-t}{10}}$ where N is the number of grams of the radioactive material and t is the time in years.
a Copy and complete the table of values for $N = 70 \times 2^{\frac{-t}{10}}$

t	0	10	20	30	40	50	60	70	80
N	70		17.5			2.2			0.27

b Draw the graph of $N = 70 \times 2^{\frac{-t}{10}}$ to show the decay of the material over 80 years.
c Use your graph to estimate the amount of radioactive material remaining after 35 years.
d A different radioactive material decays according to the formula $N = 70 \times 2^{\frac{-t}{20}}$
How would the graph of $N = 70 \times 2^{\frac{-t}{20}}$ compare with the graph of $N = 70 \times 2^{\frac{-t}{10}}$?

5 **Reasoning** The sketch shows a curve with equation $y = ka^x$ where k and a are constants and $a > 0$.
The curve passes through the points (2, 43.2) and (4, 388.8).
Work out the values of a and k.

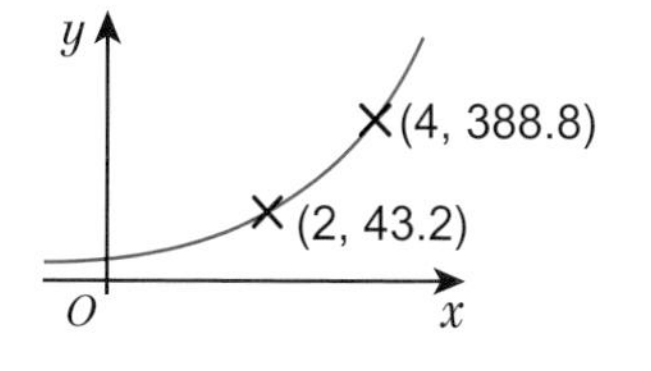

6 **Problem-solving** Here is the graph of $y = 1.5^{-x}$

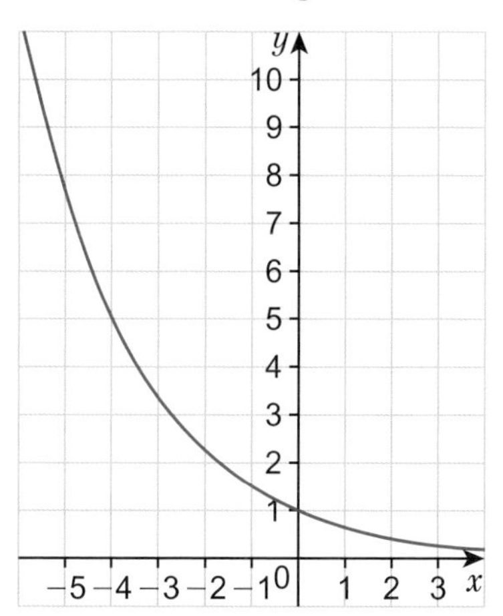

Use the graph to solve the equation $1.5^{-x} + x - 2 = 0$

7 **Problem-solving** A fishing lake is stocked with grayling.
The number of grayling in the lake T weeks after the first fish were put in is modelled by the formula $N = 120 \times 0.75^T$.

a How many grayling were put into the lake initially?
b Draw a graph to show the number of grayling in the lake for the first 8 weeks.
c Estimate the number of grayling remaining after 3 weeks.
d Estimate how long it will take for the number of grayling to reduce by 80%.
Give your answer in days.

3.5 Mixed exercise

Objective

- Consolidate your learning with more practice.

1 **Exam question**

The diagram shows a solid triangular prism.

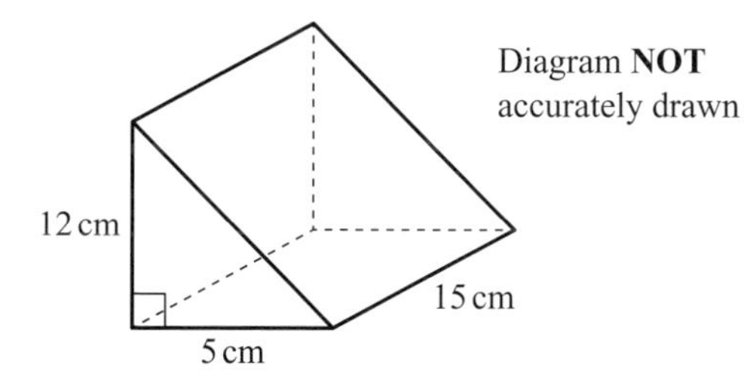

The prism is made from metal.
The density of the metal is 6.6 grams per cm^3.
Calculate the mass of the prism. **(3 marks)**

November 2012, Q13, 1MA0/2H

2 Tom has a mass of 82 kg. He stands on a stool of mass 4 kg. The stool has four legs, each with a square cross-section measuring 2.5 cm by 2.5 cm.
Work out the pressure on the floor, assuming all four feet of the stool's legs remain in contact with the floor.

Q2 hint Use $F = mg$ (where $g = 9.8\text{ m/s}^2$) to work out the total force. Remember to include the mass of the stool.

3 Jo runs 10.3 km (correct to 1 d.p.) in 44 minutes (correct to the nearest minute).
Work out the largest and smallest values of her average speed in km/h.
Give your answers correct to 1 d.p.

Q3 hint Don't round values early in a calculation as it can introduce inaccuracies in subsequent working.

4 y is directly proportional to x^2.
$y = 72$ when $x = 4$
 a Express y in terms of x
 b Work out y when $x = 6$
 c Work out x when $y = 364.5$

Q4 hint Rewrite your formula after you have found the constant of proportionality. For example, if $k = 1.4$ then $y \propto x^2$ becomes $y = 1.4x^2$.

5 **Problem-solving** Harry is making a model boat to a scale of 1:40
He uses 28 ml of paint to paint the hull of his model boat.
Estimate how much paint would be needed to paint the hull of the real boat.
Give your answer in litres.

Q5 A common error is to forget to adapt the scale factor when working out the area.

6 a Copy and complete the table of values for $y = (1.8)^{-x}$

x	−3	−2	−1	0	1	2	3
y		3.24			0.56		0.17

 b Draw the graph of $y = (1.8)^{-x}$ for $-3 \leqslant x \leqslant 3$
 c Use your graph to estimate the value of y when $x = -2.5$
 d Use your graph to estimate the value of x when $y = 1.5$

7 **Exam question**

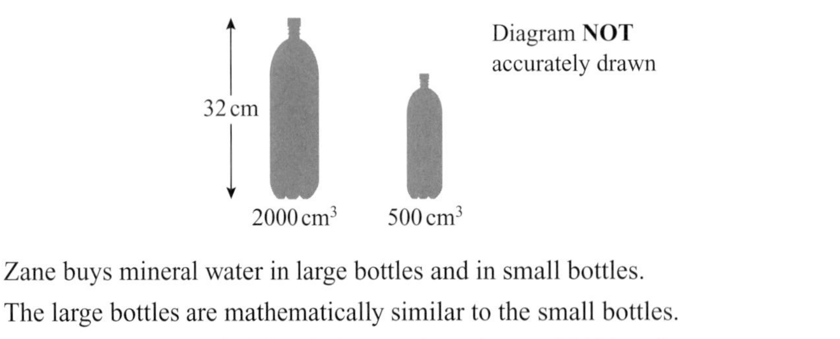

Zane buys mineral water in large bottles and in small bottles.
The large bottles are mathematically similar to the small bottles.
Large bottles have a height of 32 cm and a volume of 2000 cm^3.
Small bottles have a volume of 500 cm^3.
Work out the height of a small bottle.
Give your answer correct to 3 significant figures. **(3 marks)**

January 2015, Q15, 4MA0/3H

8 **Exam question**

P is directly proportional to q^3 $P = 270$ when $q = 7.5$
 a Find a formula for P in terms of q **(3 marks)**
 b Work out the positive value of q when $P = q$ **(2 marks)**

January 2015, Q17, 4MA0/3H

9 **Reasoning** y is inversely proportional to x^2 with constant of proportionality k.
Which of these statements is true?

A x is inversely proportional to y^2 with constant of proportionality k

B x is inversely proportional to y^2 with constant of proportionality $\frac{1}{k}$

C x is inversely proportional to $\sqrt{y}$ with constant of proportionality $\sqrt{k}$

D x is inversely proportional to $\sqrt{y}$ with constant of proportionality $\frac{1}{\sqrt{k}}$

Show working to justify your answer.

10 **Problem-solving** Two cylindrical tins of paint are mathematically similar.
The area of the base of the smaller tin is 60 cm^2.
The area of the base of the larger tin is 290.4 cm^2.
The smaller tin contains 500 ml of paint.
Work out the volume of paint in the larger tin.
Give your answer in litres.

Q10 A common error is to view the relationship between the shapes as involving addition or subtraction rather than a multiplicative relationship.

11 **Problem-solving** A yeast culture grows by 9% every hour.
At the beginning, the yeast culture has a mass of 4.5 g.

a Work out a formula for the mass, M, at a time t hours after the beginning of the experiment.

b Draw a graph to show the growth of the yeast culture over the first 10 hours.

c Estimate the mass of the yeast culture 8 hours and 45 minutes after the beginning of the experiment.
Give your answer correct to 1 d.p.

12 **Exam question**

a Sketch the graph of

$y = 3^x, x \in \mathbb{R}$,

showing the coordinates of any points at which the graph crosses the axes. **(2 marks)**

May 2014, Q8 (part a), Core 2, 1405

Q12 hint $x \in \mathbb{R}$ means that x is a real number (an integer, a fraction or a surd).

13 Two variables, m and t, are connected by the relationship $m \propto \frac{1}{\sqrt{t}}$
When $m = 6.75$, $t = 16$
Work out values for m and t that satisfy

a $m = t$

b $t = 4m$

Give your answers correct to 3 s.f. where appropriate.

4 PROBABILITY

What is covered in this section	Higher Student Book reference
4.1 Independent events and tree diagrams • Draw and use tree diagrams to work out probabilities of independent events.	Unit 10
4.2 Conditional probability • Decide if events are dependent or independent. • Draw and use tree diagrams to work out conditional probabilities of dependent events.	Unit 10
4.3 Venn diagrams and set notation • Use Venn diagrams to calculate probabilities. • Use set notation to describe regions of Venn diagrams.	Unit 10
4.4 Mixed exercise • Consolidate your learning with more practice.	

4.1 Independent events and tree diagrams

Objective

- Draw and use tree diagrams to work out probabilities of independent events.

Key point 1

Events are **independent** when the outcome of one does not affect the outcome of the other
To find the combined probability of two independent events, multiply the probabilities.
P(A and B) = P(A) × P(B)

Key point 2

A **tree diagram** shows the possible outcomes of two or more combined events.

Warm up

1 A bag contains 4 red (R) discs and 3 blue (B) discs.
One disc is taken at random, its colour is noted and then it is replaced.
A second disc is then taken.

a Copy and complete this tree diagram showing the possible outcomes.

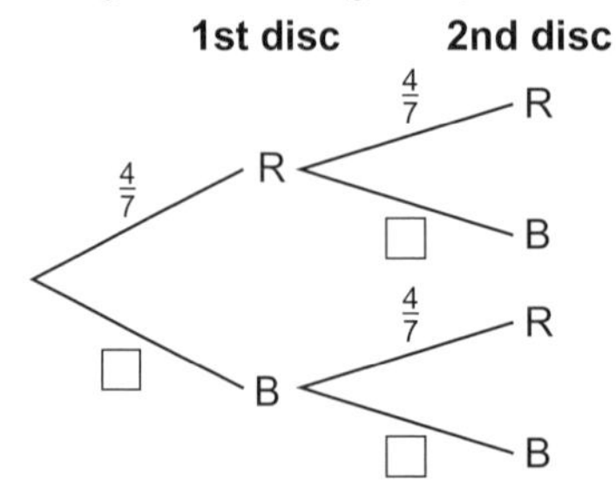

b Work out the probability of taking two blue discs.

2 A fair five-sided spinner has two green (G) sections and three yellow (Y) sections.
The spinner is spun twice and the result of each spin is recorded.

a Draw a tree diagram to show all the possible outcomes.

b Work out the probability of both spins being the same colour.

3 **Reasoning** Tom and Jerry are having trouble with their tree diagram probability questions.
The probability that Tom gets a question correct is 0.6
The probability that Jerry gets a question correct is 0.45
They both attempt another question.
What is the probability that only one of them gets it correct?

Q3 hint Draw a tree diagram to show all the possible outcomes.

4 **Reasoning** A bag contains 4 red, 3 blue and 2 green counters.
A counter is taken at random, its colour is noted and then it is replaced.
A second counter is then taken.
Work out the probability that

a the two counters are the same colour
b only one of the counters is blue
c at least one of the counters is green.

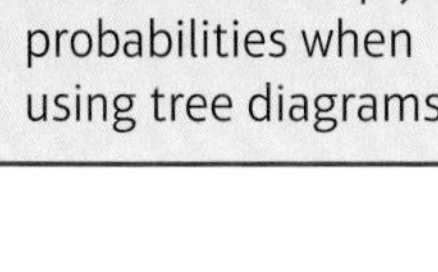

5 **Reasoning** Alice, Beth and Carla are taking their driving test.
The probability that Alice will pass is 0.85
The probability that Beth will pass is 0.75
The probability that Carla will pass is 0.8

a Draw a tree diagram to show all the possible outcomes for the three women.
b Work out the probability that all three pass the test.
c Work out the probability that all three fail the test.
d Work out the probability that at least one of them fails the test.
e Work out the probability that only one of them fails.

Q5a hint The first-stage branches will be 'Alice passes' and 'Alice fails'.

6 **Problem-solving** A fairground game involves rolling a fair 1–6 dice three times.
If you roll three 6s you win £20.
If you roll two 6s you win £10.
The stall-holder charges people £2 to play the game.
432 people play the game.
Estimate the stall-holder's profit.

4.2 Conditional probability

Objectives

- Decide if events are dependent or independent.
- Draw and use tree diagrams to work out conditional probabilities of dependent events.

Key point 3

When events are **dependent**, the outcome of the first event influences the outcome of the second event.
Conditional probabilities are probabilities associated with dependent events.

Key point 4

In a tree diagram showing conditional probabilities, the probabilities on the branches for the second outcome depend on what has happened in the first outcome.
The probabilities on the sets of branches for the second outcome are different from each other.

Warm up

1 Matt has 4 red (R) socks and 6 black (B) socks in a drawer.
He takes out two socks at random.

a Copy and complete this tree diagram to show the possible outcomes.

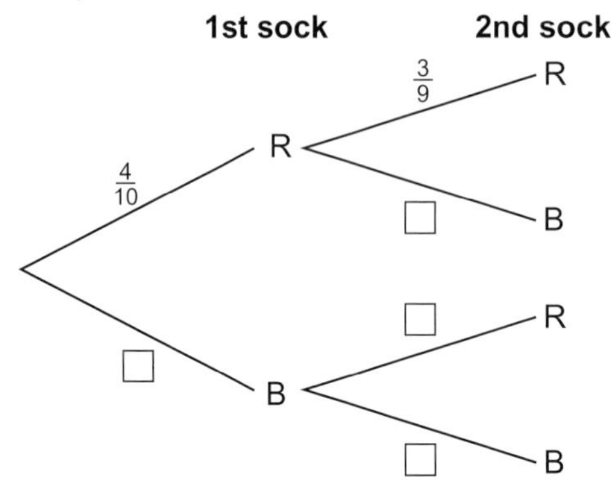

b Work out the probability that Matt takes a matching pair of socks.

Q1a hint When two items are taken at the same time, it is the same as if one item is taken, but not replaced, and then another item is taken. The second outcome depends on the first outcome.

Q1 A common error is to forget that once one sock has been taken out, there are fewer socks in the drawer for the second pick.

2 **Reasoning** A bag contains 7 toffees and 5 fruit sweets.
Harry takes a sweet at random and eats it.
He then eats a second sweet.

a Work out the probability that he eats two toffees.

b Work out the probability that he eats at least one toffee.

Q2 hint Draw a tree diagram. The probabilities are conditional probabilities.

3 **Reasoning** Dan has to attend a meeting.
The probabilities of dry weather (D), rain (R) or snow (S) are
P(D) = 0.5 P(R) = 0.3 P(S) = 0.2
If it is dry the probability that Dan will arrive in time for the meeting is 0.8
If it rains the probability that he will arrive in time for the meeting is 0.4
If it snows the probability that he will arrive in time for the meeting is 0.15
Is he more likely to arrive in time for the meeting or to be late?

4 **Reasoning** Andy and Roger play a tennis match.
The probability that Andy wins the first set is 0.55
When Andy wins a set, the probability that he wins the next set is 0.65
When Roger wins a set, the probability that he wins the next set is 0.6
The first person to win two sets wins the match.

a Draw a tree diagram to show all the possible outcomes.

b Calculate the probability that Andy wins the match by 2 sets to 1.

c Calculate the probability that Roger wins the match.

Q4 hint Show your working, including when you intend to multiply or add probabilities.

5 **Reasoning** A bag contains some coloured counters.
Three are red (R), five are green (G) and two are yellow (Y).

Two counters are taken from this bag at random.
Work out the probability that

a both counters are yellow

b at least one counter is red

c the two counters are different colours.

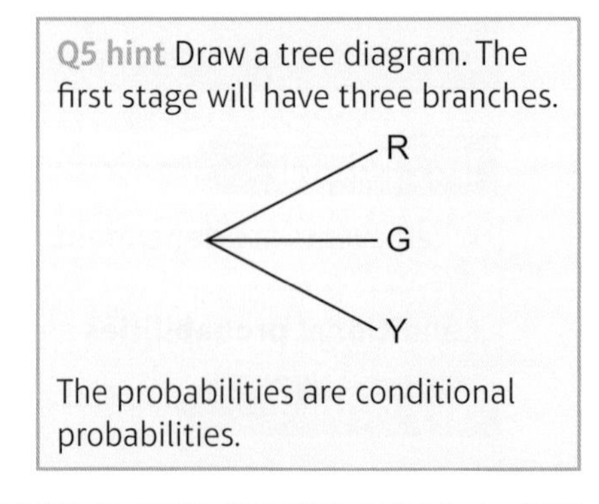

Q5c hint Think of a quick way to do this. What is the opposite of 'different colours'?

6 **Reasoning** To complete a training programme successfully, candidates must pass a test which they take at the end of it.
If a candidates fails the first time, they re-sit the test.
The probability of passing the test at the first attempt is 0.7
Those who fail are allowed only one re-sit.
The probability of passing the re-sit is 0.4

a Complete the tree diagram, which shows all the possible outcomes.

b What is the probability that a candidate passes the test and successfully completes the training programme?

c Tim and Louise both take the training programme.
What is the probability that only one of them is successful?

1st attempt | Re-sit
0.7
☐ Fail
0.4 Pass
☐ Fail

Q6c hint Think about how many stages your tree diagram will need.

7 **Reasoning** Jon has 8 tins of soup in his cupboard, but all the labels are missing.
He knows that there are 5 tins of tomato soup and 3 tins of mushroom soup.
He opens three tins at random.
Work out the probability that

a all three tins are the same variety of soup

b he opens more tins of mushroom soup than tomato soup.

8 **Problem-solving** Bag A contains 4 red and 3 green counters.
Bag B contains 2 red and 5 green counters.
Move 1 A counter is taken from bag A and placed into bag B.
Move 2 A counter is taken from bag B and placed into bag A.
Work out the probability that bag A has more red counters than green counters after these two moves.

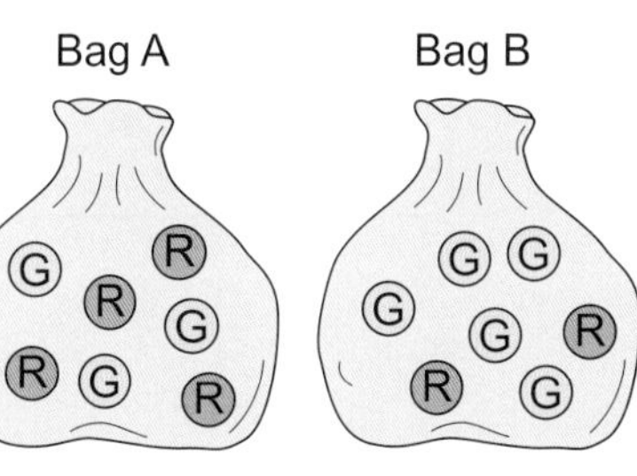

4.3 Venn diagrams and set notation

Objectives

- Use Venn diagrams to calculate probabilities.
- Use set notation to describe regions of Venn diagrams.

Key point 5

A ∩ B means 'A intersection B'.
This is all the elements that are in A *and* in B.

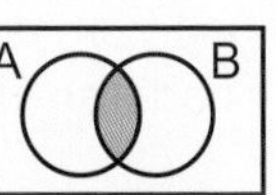
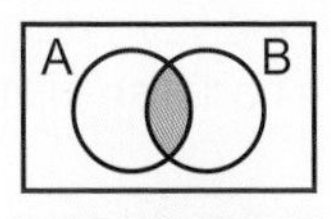

A ∪ B means 'A union B'.
This is all the elements that are in A *or* B *or* both.

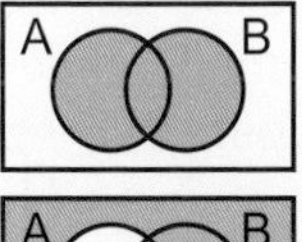

A′ means the elements *not* in A.

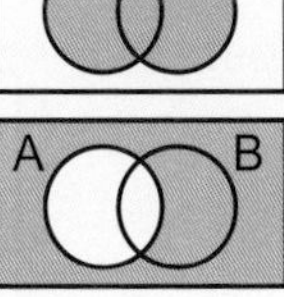

ξ means the universal set – all the elements being considered.

Key point 6

A ∩ B ∩ C means the **intersection** of A, B and C.
A ∪ B ∪ C means the **union** of A, B and C.
P(A ∩ B | B) means the probability of A and B given B.

Warm up

1 Lily carried out a survey of 92 students to find out how many had a holiday at home (H) and how many had a holiday abroad (A) this year.

- 35 students had a holiday at home.
- 11 students had a holiday at home and a holiday abroad.
- 19 students did not have a holiday at all.

a Draw a Venn diagram to show Lily's data.

A student is picked at random from this group.

Work out the probability that they

b went on holiday abroad

c went on holiday abroad, given that they had a holiday at home.

Q1c hint How many had a holiday at home? How many of these also had a holiday abroad? P(A|H) = $\frac{\square}{\square}$

2 The Venn diagram shows the numbers of students who take maths (M), English (E) and history (H).

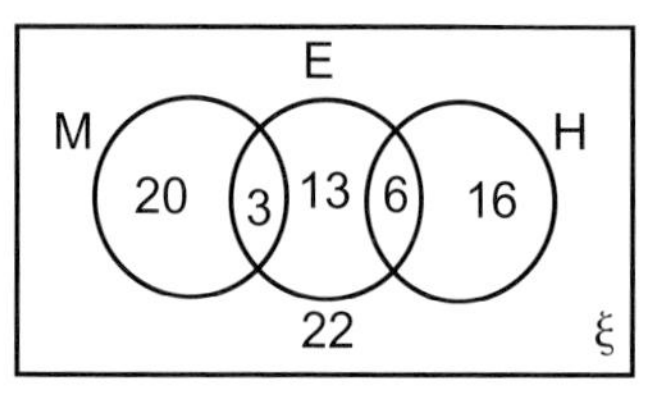

Work out

a P(M ∪ E)

b P(E ∩ H | E)

Q2b hint P(E ∩ H | E) means the probability of a student taking English and history, given that they take English.

3 **Reasoning** 150 people were asked which of the countries France, Holland and Spain they had visited.

- 80 people had been to France, 52 to Holland and 63 to Spain.
- 21 people had been to France and Holland.
- 28 people had been to France and Spain.
- 25 people had been to Holland and Spain.
- 12 people had been to all three countries.

a Draw a Venn diagram to represent this information.

b Work out the probability that a person, picked from this group at random, had visited only two of the three countries.

c Given that a person had visited Spain, work out the probability that they had also visited France.

Q3a hint First of all, put 12 in the overlap of all three regions of your Venn diagram.

4 The Venn diagram shows the ice-cream flavours chosen by a group of 44 children at a party. The choices are strawberry (S), choc-chip (C) and toffee (T).
A child is picked at random.
Work out

a P(S)

b P [(S ∪ C ∪ T)′]

c P(T ∪ C | C)

d P(C | S ∪ T)

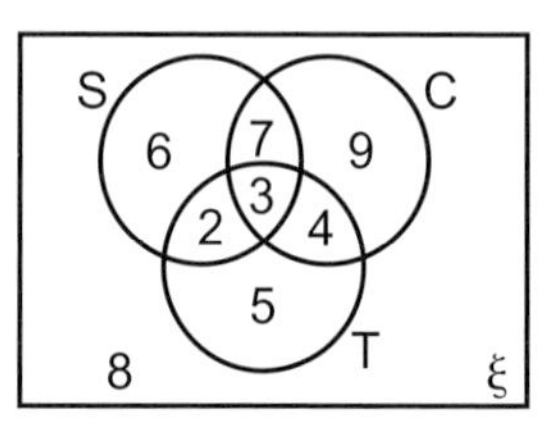

Q4 hint Check the set notation – see Key points 5 and 6.

5 The Venn diagram shows the sports played by a group of 80 students.
The three most popular sports were tennis (T), golf (G) and cricket (C).
A student is picked at random.
Work out

a $P(T \cap C \cap G')$

b $P(T \cap G \cap C \mid T)$

c $P(T \mid G \cup C)$

d $P(G \mid C')$

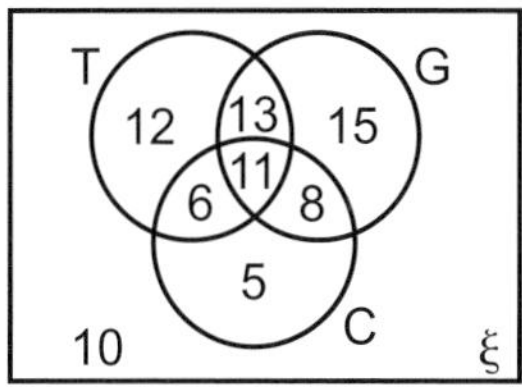

Q5 hint Check the set notation – see Key points 5 and 6.

6 **Reasoning** A gym offers three different classes: Total Spin (T), Bootcamp (B) and Zumba (Z).
70 members of the gym were asked which of these classes they had attended.

- 24 people had attended Total Spin.
- 28 people had attended Bootcamp.
- 30 people had attended Zumba.
- 10 people had attended Total Spin and Bootcamp.
- 12 people had attended Bootcamp and Zumba.
- 7 people had attended Total Spin and Zumba.
- 4 people had attended all three classes.

a Draw a Venn diagram to represent this information.

A person is picked at random.

b Work out i $P(T \cup B \cup Z)$ ii $P(Z \mid B \cup T)$ iii $P(B \mid Z')$

4.4 Mixed exercise

Objective

- Consolidate your learning with more practice.

1 **Reasoning** There are 31 students in a class.
The only languages available for the class to study are French and Spanish.

- 17 students study French.
- 15 students study Spanish.
- 6 students study neither French nor Spanish.

a Draw a Venn diagram to represent this information.

b Work out i $P(F \cap S)$ ii $P(S \mid F')$

2 **Reasoning** 75 children were asked whether they had a cat (C), a dog (D) or a rabbit (R) as a pet.

- 32 children had a cat, 36 had a dog and 18 had a rabbit.
- 13 children had a cat and a dog.
- 9 children had a cat and a rabbit.
- 5 children had a dog and a rabbit.
- 4 children had all three pets.

a Draw a Venn diagram to represent this information.

b Work out i $P(C \cap R \cap D')$ ii $P(C \mid D \cup R)$ iii $P(D \mid C')$

3 **Reasoning** A machine is manufacturing components.
The probability of a component being faulty is 4%.
Three components are made.
Work out the probability that

a all three components are perfect

b only one component is faulty

c at least one component is faulty.

Q3 Drawing a tree diagram is often a key to success. Students who attempt a question without drawing one often make mistakes.

4 **Exam question**

Bill and Jo play some games of table tennis.

The probability that Bill wins the first game is 0.7

When Bill wins a game, the probability that he wins the next game is 0.8

When Jo wins a game, the probability that she wins the next game is 0.5

The first person to win two games wins the match.

a Complete the probability tree diagram.

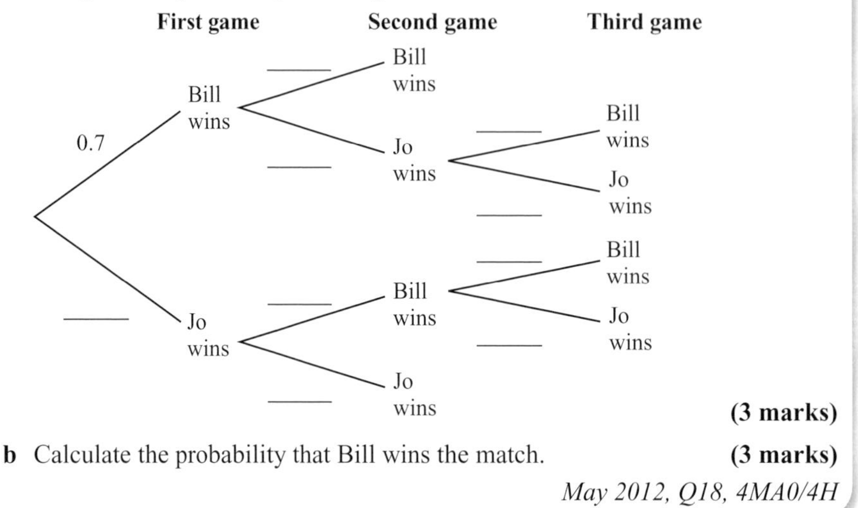

(3 marks)

b Calculate the probability that Bill wins the match. **(3 marks)**

May 2012, Q18, 4MA0/4H

5 **Reasoning** A variety pack of yoghurts contains 3 strawberry, 5 black cherry and 4 pineapple.

Penny eats two yoghurts from this pack, picking them at random.

Work out the probability that

a both yoghurts are the same flavour

b at least one is strawberry

c only one of them is black cherry.

6 **Exam question**

Peter wants to pass his driving test.

The probability that he passes at his first attempt is 0.7

When Peter passes his driving test, he does not take it again.

If he fails, the probability that he passes at the next attempt is 0.8

a Complete the probability tree diagram for Peter's first two attempts.

First attempt **Second attempt**

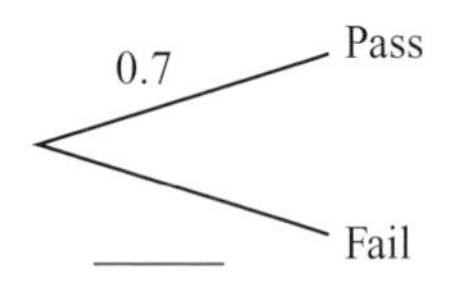

(2 marks)

b Calculate the probability that Peter needs exactly two attempts to pass his driving test. **(2 marks)**

c Calculate the probability that Peter passes his driving test at his third or fourth attempt. **(3 marks)**

January 2014, Q14, 4MA0/4H

5 GEOMETRY AND MEASURES

What is covered in this section	Higher Student Book reference
5.1 Transformations • Combine the transformations of translation, reflection, rotation and enlargement (including fractional and negative scale factors) and describe the single transformation that maps any one shape onto another. • Understand the changes and invariances achieved by combinations of transformations.	Unit 8
5.2 Area and volume • Calculate arc lengths, angles and areas of sectors and segments of circles. • Calculate surface areas and volumes of pyramids, cones, frustums and spheres.	Units 7, 13
5.3 Circle theorems • Recall circle properties. • Recall proofs of circle theorems. • Use circle theorems and circle properties to solve geometrical problems.	Unit 16
5.4 Trigonometry • Know the exact values of the trigonometric functions for angles of 0°, 30°, 45°, 60° and 90°. • Use trigonometric formulae to solve problems in 2D and 3D, including finding the angle between a line and a plane.	Units 5, 13
5.5 Vectors • Apply addition and subtraction of vectors, multiplication by a scalar, and diagrammatic and column representation of vectors to solve problems in 2D. • Use vector methods to prove that lines are parallel or that points are collinear, and to construct geometric arguments and proofs.	Unit 18
5.6 Mixed exercise • Consolidate your learning with more practice.	

5.1 Transformations

Objectives

- Combine the transformations of translation, reflection, rotation and enlargement (including fractional and negative scale factors) and describe the single transformation that maps any one shape onto another.
- Understand the changes and invariances achieved by combinations of transformations.

Key point 1

In a **translation**, all the points on the shape move the same distance in the same direction.
You can describe a translation by using a **column vector**.
The column vector for a translation 2 squares right and 3 squares down is $\begin{pmatrix} 2 \\ -3 \end{pmatrix}$.
The top number gives the movement parallel to the x-axis.
The bottom number gives the movement parallel to the y-axis.

Key point 2

To describe a **reflection** you need to give the equation of the mirror line. The object and image are the same perpendicular distance from the mirror line but on opposite sides.

Key point 3

To describe a **rotation** you need to give

- the direction of turn (clockwise or anticlockwise)
- the angle of turn
- the centre of rotation.

Key point 4

After a translation, reflection or rotation, the object and image are **congruent**.

Key point 5

An **enlargement** is a transformation where all the side lengths of a shape are multiplied by the same **scale factor**.
To describe an enlargement you need to give the centre of enlargement and the scale factor.

Key point 6

When the scale factor of an enlargement is a fraction (that is, is less than 1), the image is smaller than the object.
When the scale factor is negative, the image is on the opposite side of the centre of enlargement and appears upside down. Its size is determined by the scale factor.
In this diagram A is the object.

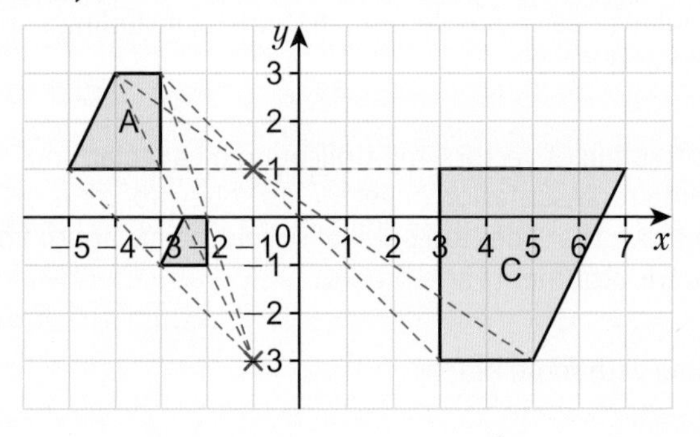

B is the image of A after an enlargement of scale factor $\frac{1}{2}$, centre of enlargement (–1, –3).
C is the image of A after an enlargement of scale factor –2, centre of enlargement (–1, 1).

Warm up

1 Draw a coordinate grid from –6 to 6 on both axes.
Draw triangle A at (2, 1), (2, 2) and (4, 2).

a Translate A by the vector $\begin{pmatrix} -2 \\ 4 \end{pmatrix}$. Label the image B.

b Reflect A in the line $y = -1$. Label the image C.

c Reflect A in the line $y = x$. Label the image D.

d Rotate A 180° about (0, –2). Label the image E.

e Rotate A 90° anticlockwise about (–1, –2). Label the image F.

f Describe the single transformation that maps B onto E.

g Describe the single transformation that maps D onto C.

2 Draw a coordinate grid from –6 to 6 on both axes.
Draw triangle J at (–2, 3), (2, 3) and (1, 5).

a Translate J by the vector $\begin{pmatrix} 3 \\ -3 \end{pmatrix}$. Label the image K.

b Reflect K in the line $y = -x$. Label the image L.

c Reflect L in the line $x = -3$. Label the image M.

d Describe the single transformation that maps M onto J.

3 Draw a coordinate grid from –6 to 6 on both axes.
Draw trapezium P at (–4, –1), (–6, –1), (–6, –4) and (–4, –2).

a Reflect P in the line $y = x$. Label the image Q.
b Rotate Q 90° anticlockwise about (0, –1). Label the image R.
c Translate R by the vector $\begin{pmatrix} -4 \\ 6 \end{pmatrix}$. Label the image S.
d Rotate S 90° anticlockwise about (1, 6). Label the image T.
e Reflect T in the line $y = x$. Label the image U.
f Describe the single transformation that maps U onto P.
g Describe the single transformation that maps S onto Q.
h What is true about the areas of all the trapezia?

4 Draw a coordinate grid from –6 to 6 on both axes.
Draw triangle A at (–2, –1), (–1, –1) and (–1, 1).

a Rotate A 90° clockwise about (1, 2). Label the image B.
b Enlarge B by a scale factor of 2, centre of enlargement (–4, 6). Label your image C.
c Translate C by the vector $\begin{pmatrix} 2 \\ -6 \end{pmatrix}$. Label the image D.
d Rotate D 90° anticlockwise about (–3, –5). Label the image E.
e Describe the single transformation that maps E onto A.
f What is the ratio of the area of triangle E to the area of triangle A?

5 Draw a coordinate grid from –8 to 8 on both axes.
Draw triangle ABC at A (–3, 8), B (–5, 4) and C (–1, 6).

a Enlarge triangle ABC by a scale factor of –2, centre of enlargement (–1, 3).
Label the image PQR, where A maps onto P, B maps onto Q and C maps onto R.
b What do you notice about the corresponding sides of the object and image?
c What enlargement maps triangle PQR onto triangle ABC?
d What is the ratio of the area of triangle ABC to the area of triangle PQR?
e Enlarge triangle ABC by a scale factor of $2\frac{1}{2}$, centre of enlargement (–5, 8).
Label the image STU, where A maps onto S, B maps onto T and C maps onto U.
f What is the ratio of the lengths of the sides of triangle STU to those of triangle PQR?
g What is the ratio of the area of triangle STU to the area of triangle PQR?

6 Draw a coordinate grid from –8 to 8 on both axes.
Draw your answers to parts **a–d** on the grid.

a Draw triangle A at (1, 2), (1, –4) and (3, –4), and enlarge it by a scale factor of $-\frac{1}{2}$, centre of enlargement (–1, –2).
Label the image E.
b Draw triangle B at (2, 2), (5, 2) and (4, 4), and enlarge it by a scale factor of $-1\frac{1}{2}$, centre of enlargement (0, 0).
Label the image F.
c Draw trapezium C at (–6, 0), (–6, –2), (–4, –2) and (–4, 2), and enlarge it by a scale factor of $1\frac{1}{2}$, centre of enlargement (–8, –6).
Label the image G.
d Draw triangle D at (0, 4), (8, 4) and (8, 8), and enlarge it by a scale factor of $-\frac{1}{2}$, centre of enlargement (4, 5).
Label the image H.

Q6a A common error is to enlarge by a scale factor of $\frac{1}{2}$ instead of $-\frac{1}{2}$.

5.2 Area and volume

Objectives

- Calculate arc lengths, angles and areas of sectors and segments of circles.
- Calculate surface areas and volumes of pyramids, cones, frustums and spheres.

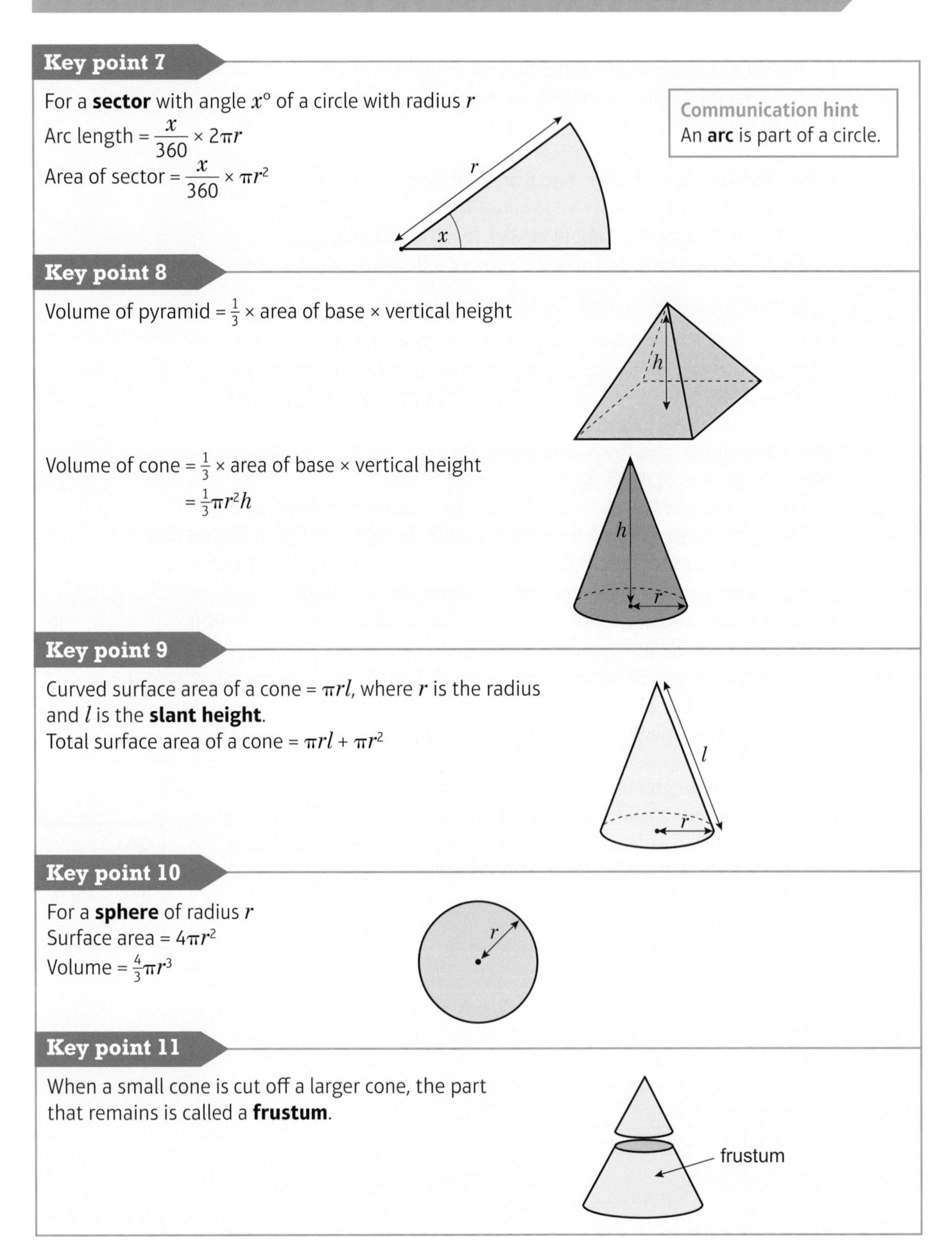

Key point 7

For a **sector** with angle $x°$ of a circle with radius r

Arc length $= \frac{x}{360} \times 2\pi r$

Area of sector $= \frac{x}{360} \times \pi r^2$

Communication hint
An **arc** is part of a circle.

Key point 8

Volume of pyramid $= \frac{1}{3} \times$ area of base $\times$ vertical height

Volume of cone $= \frac{1}{3} \times$ area of base $\times$ vertical height

$= \frac{1}{3}\pi r^2 h$

Key point 9

Curved surface area of a cone $= \pi r l$, where r is the radius and l is the **slant height**.

Total surface area of a cone $= \pi r l + \pi r^2$

Key point 10

For a **sphere** of radius r

Surface area $= 4\pi r^2$

Volume $= \frac{4}{3}\pi r^3$

Key point 11

When a small cone is cut off a larger cone, the part that remains is called a **frustum**.

1 Work out the diameter of a circle of area 45 cm^2.
Give your answer correct to 3 s.f.

2 Work out the curved surface area and the volume of a cylindrical rod of length 1.5 m and diameter 2.4 cm.
Give your answers correct to 3 s.f.

3 Work out the arc length and area of each sector.
Give your answers correct to 3 s.f.

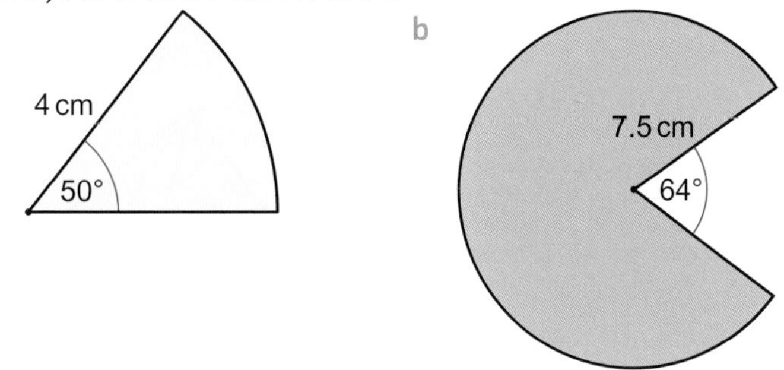

4 This circle has a circumference of length 30π cm.
The length of the arc from A to B is 6π cm.

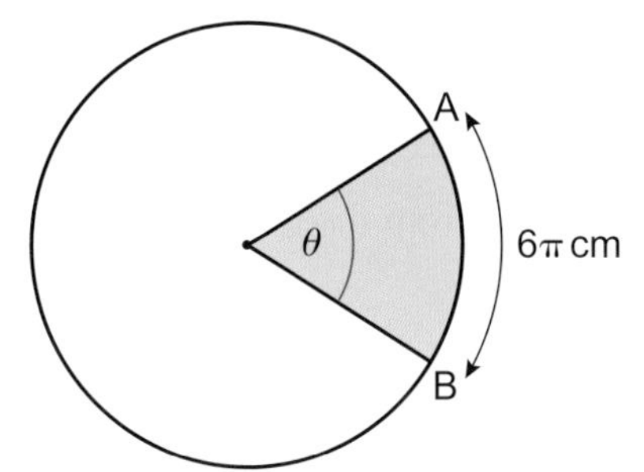

Work out

a the value of θ

b the radius of the circle

c the area of the minor sector.

Leave your answer to part **c** in terms of π.

5 This circle has a radius of 8 cm, and the length of the arc from P to Q is 15 cm.

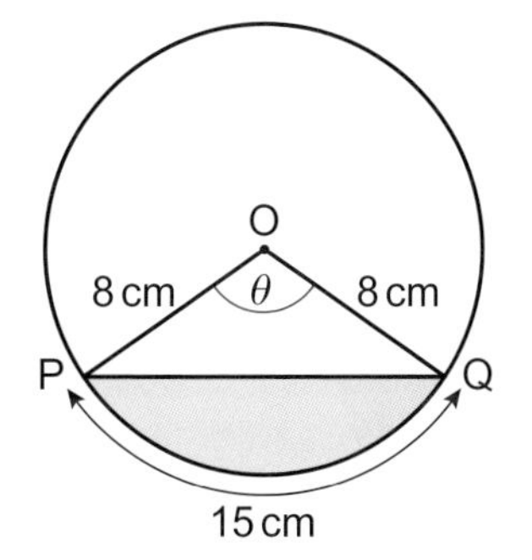

Work out

a the value of θ

b the length of the chord PQ

c the area of the minor segment (shaded).

Give all your answers correct to 3 s.f.

Q5b hint Triangle POQ is isosceles. Draw a line from O to the midpoint of PQ. This line is perpendicular to the chord. Use the exact value of θ in your calculations.

Q5c hint Area of the segment
= area of sector POQ – area of triangle POQ
Use area = $\frac{1}{2}ab \sin C$ for the area of triangle POQ.

Warm up

6 Work out the volume and curved surface area of each cone. Give your answers correct to 3 s.f.

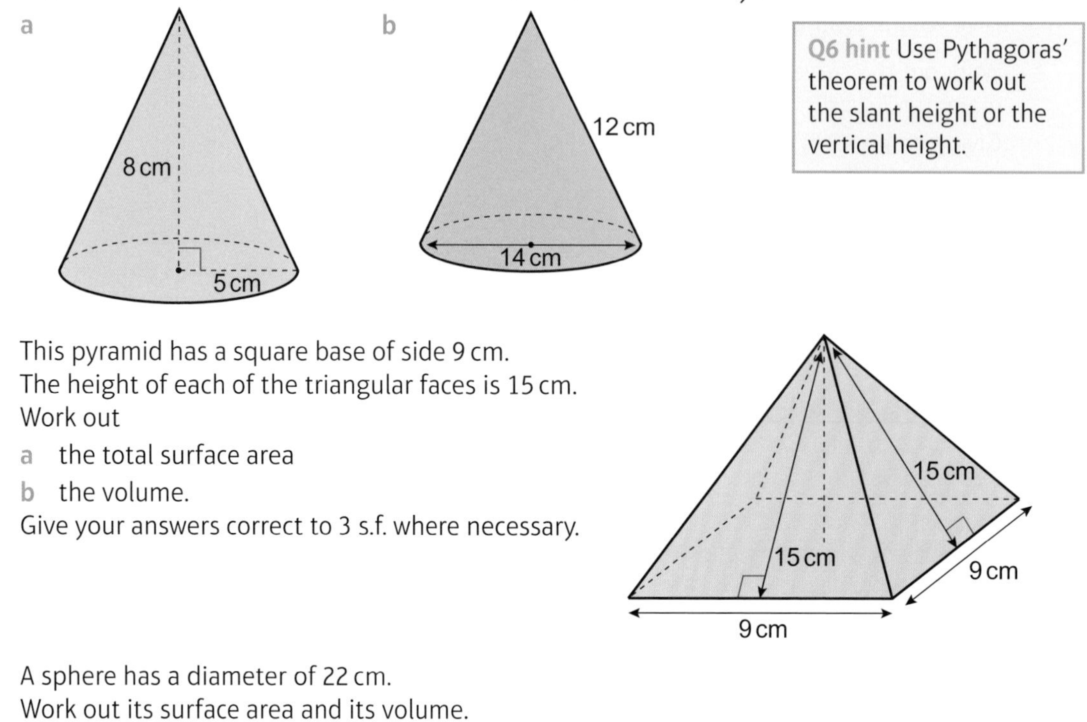

Q6 hint Use Pythagoras' theorem to work out the slant height or the vertical height.

7 This pyramid has a square base of side 9 cm.
The height of each of the triangular faces is 15 cm.
Work out

a the total surface area

b the volume.

Give your answers correct to 3 s.f. where necessary.

8 A sphere has a diameter of 22 cm.
Work out its surface area and its volume.
Give your answers correct to 3 s.f.

9 **Problem-solving** This cone is cut one-third of the way from the top.

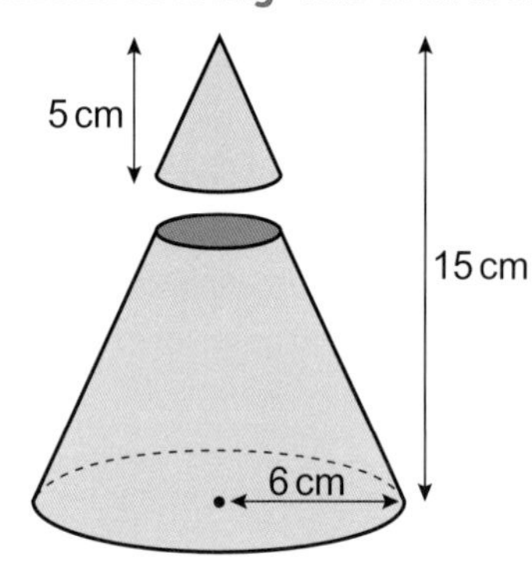

Q9 A common error is to work out the radius of the small cone incorrectly. Use similar triangles to work out the radius of the small cone.

Work out the volume of the frustum. Give your answer correct to 3 s.f.

10 **Problem-solving** A solid is made from a hemisphere with a cone fitted on top of it.
Work out the surface area and the volume of the solid.
Give your answers in terms of π.

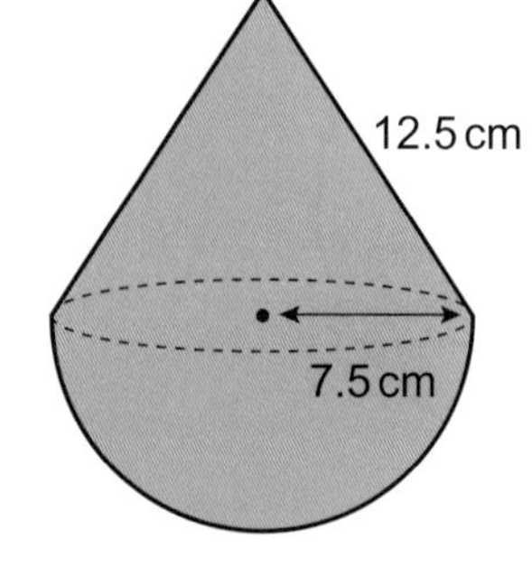

Q10 hint What do you need to know in order to calculate the volume?

11 **Problem-solving** The diagram shows a plinth for a statue. It is made by cutting off the top of a cone to make a frustum.
Work out the volume of the plinth, giving your answer correct to 3 s.f.

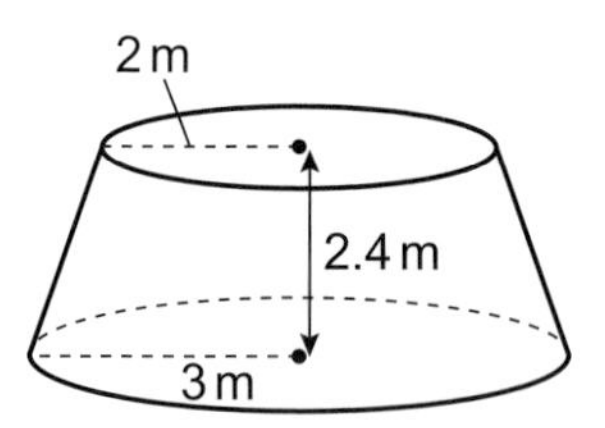

Q11 A common error is to use the wrong formula. For the volume of a cone, make sure you use $\frac{1}{3}\pi r^2 h$.

12 Problem-solving The diagrams show a sector made out of paper and a paper cup, in the shape of a cone, made by joining together the points marked A and B.

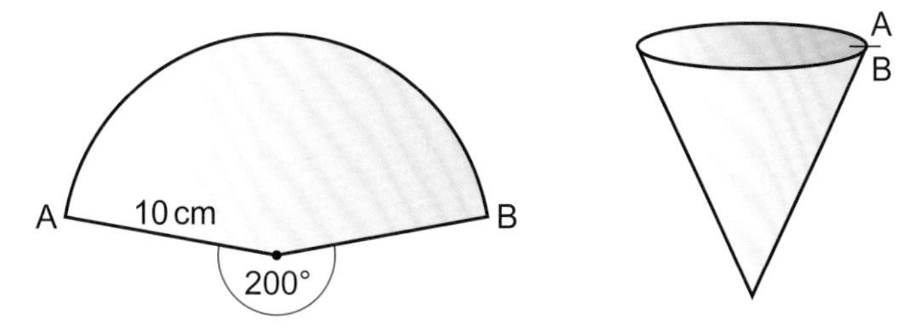

Work out the curved surface area and the capacity of the paper cup, giving your answers correct to 3 s.f.

5.3 Circle theorems

Objectives

- Recall circle properties.
- Recall proofs of circle theorems.
- Use circle theorems and circle properties to solve geometrical problems.

Key point 12

The line from the centre of a circle perpendicular to a chord bisects the chord.

Proof

Join OA and OB.
OA = OB (radii)
Angle OMA = angle OMB = 90° (given)
OM is common.
Triangles OMA and OMB are congruent (RHS).
Therefore AM = BM.

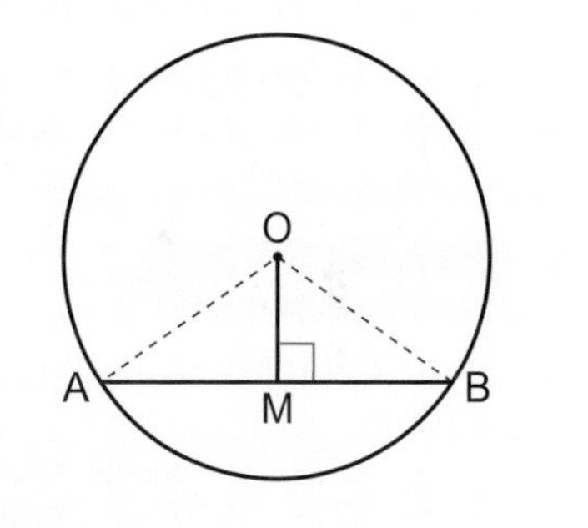

Key point 13

The angle between a tangent and a radius is a right angle.

Proof

Draw line OAB. Assume that OP is *not* perpendicular to PT and that OB *is* perpendicular to PT.
If angle OBP = 90°, then angle OPB < 90° (triangle property).
So OP > OB (greater side opposite greater angle).
But OP = OA (radii), therefore OA > OB which is clearly not true.
So the original assumption is incorrect.
It is also incorrect for all other positions of B.
Therefore OP *is* perpendicular to PT.
Therefore angle OPT = 90°.
(This is an example of **proof by contradiction**.)

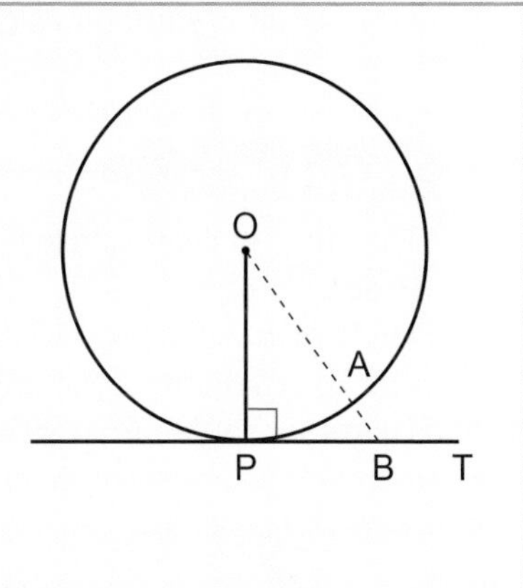

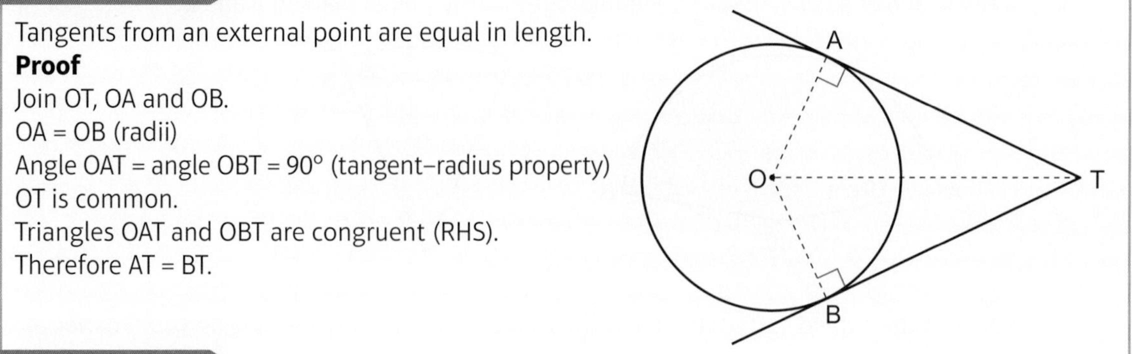

Key point 14

Tangents from an external point are equal in length.

Proof

Join OT, OA and OB.
OA = OB (radii)
Angle OAT = angle OBT = 90° (tangent–radius property)
OT is common.
Triangles OAT and OBT are congruent (RHS).
Therefore AT = BT.

Key point 15

The angle subtended by an arc at the centre of a circle = 2 × the angle that it subtends at the circumference.

Proof

Triangles OAP and OBP are isosceles.
Let angles APO and PAO = x (base angles of isosceles triangle APO).
Let angles BPO and PBO = y (base angles of isosceles triangle BPO).
Angle AOC = $2x$ (exterior angle property of triangle APO)
Angle BOC = $2y$ (exterior angle property of triangle BPO)
Angle AOB = $2x + 2y = 2(x + y)$ = 2 × angle APB

Three diagrams related to this theorem are

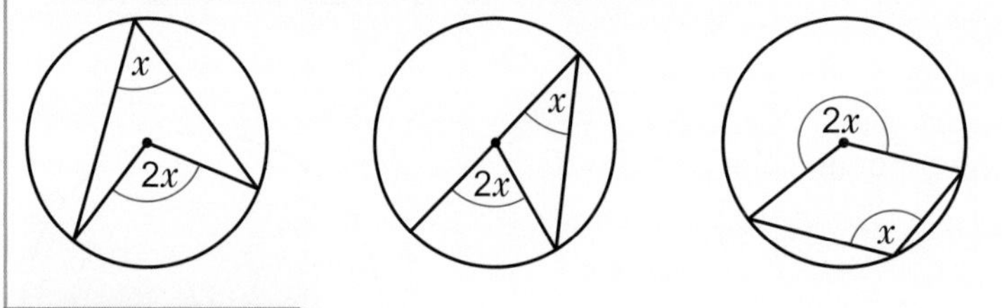

Key point 16

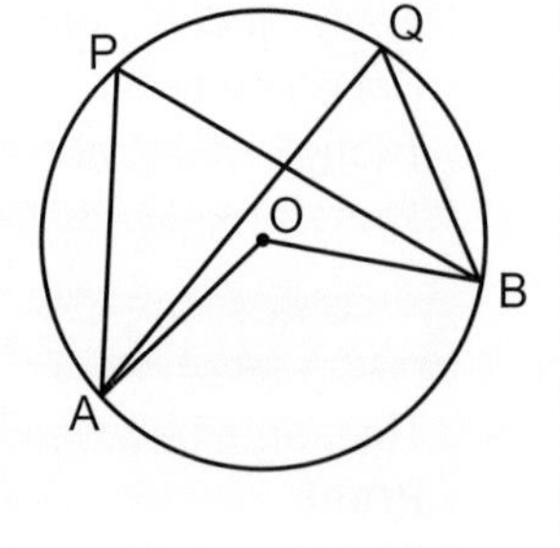

Angles subtended by the same arc are equal.

Proof

Angle AOB = 2 × angle APB (angle at centre = 2 × angle at circumference)
Angle AOB = 2 × angle AQB (angle at centre = 2 × angle at circumference)
Therefore angle APB = angle AQB.
(This result is sometimes called **angles in the same segment**.)

Key point 17

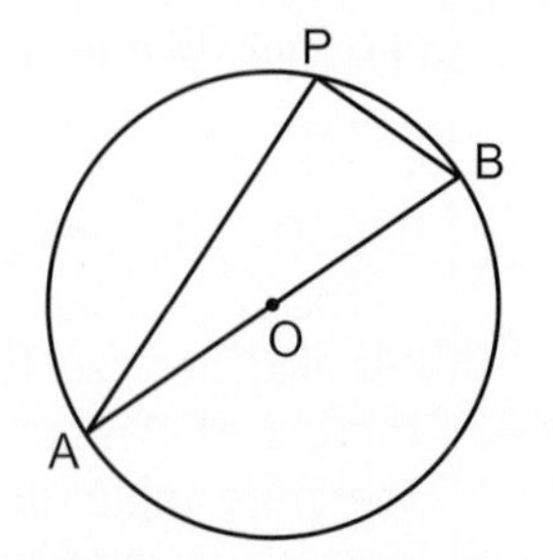

The angle in a semicircle is a right angle.

Proof

Angle AOB = 2 × angle APB (angle at centre = 2 × angle at circumference)
But AB is a diameter, so angle AOB = 180°.
Therefore angle APB = 90°.

Key point 18

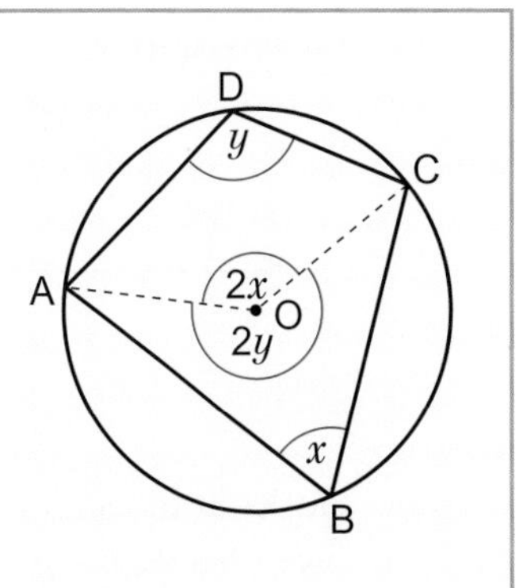

Opposite angles of a cyclic quadrilateral are supplementary (add up to 180°).

Proof

ABCD is a cyclic quadrilateral, i.e. a quadrilateral whose vertices lie on the circumference of a circle.
Join A and C to O, the centre of the circle.
Let angle ABC = x and angle ADC = y.
Angle AOC = $2x$ (angle at centre = 2 × angle at circumference)
Reflex angle AOC = $2y$ (angle at centre = 2 × angle at circumference)
Angle AOC + reflex angle AOC = $2x + 2y = 360°$ (angles round a point)
Therefore $x + y = 180°$, i.e. angle ABC + angle ADC = 180°.
(Similarly, angle DAB + angle DCB = 180°.)

Key point 19

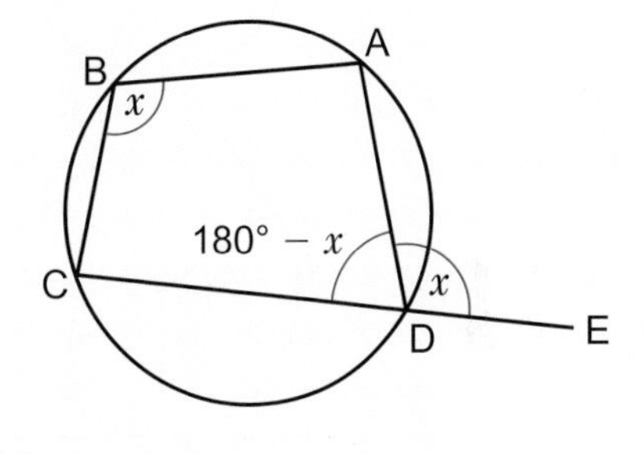

Each exterior angle of a cyclic quadrilateral is equal to the opposite interior angle.

Proof

Let angle ADE = x.
Angle ADC = 180° − x (angles on a straight line)
Angle ABC = 180° − angle ADC (opposite angles of a cyclic quadrilateral are supplementary)
Angle ABC = 180° − (180° − x) = x
Therefore angle ADE = angle ABC.

Key point 20

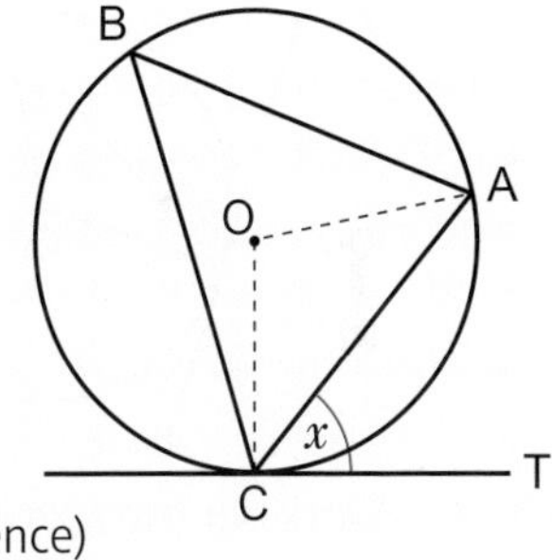

The angle between a tangent and a chord is equal to the angle in the alternate segment (i.e. the angle subtended by the chord on the circumference of the circle).

Proof

Let angle ACT = x
Angle OCA = 90° − x (angle between tangent and radius = 90°)
Angle OAC = 90° − x (base angles of isosceles triangle OAC)
Angle AOC = 180° − (90° − x) − (90° − x) = $2x$ (angle sum of triangle AOC)
Angle ABC = $\frac{1}{2}$ of angle AOC = x (angle at centre = 2 × angle at circumference)
Therefore angle ACT = angle ABC.

Three diagrams associated with this theorem are

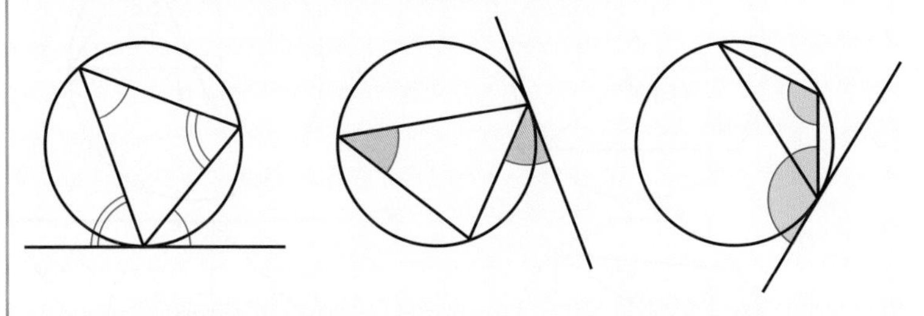

1 Work out the value of x in each of these diagrams.
Give reasons for your answers.

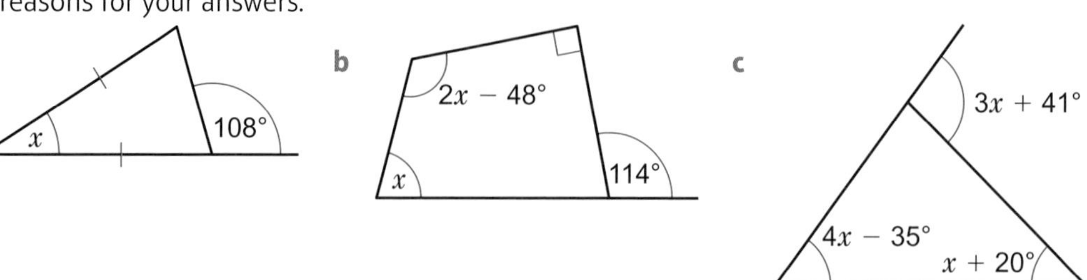

2 Work out the size of angle x in each of these diagrams.
Give reasons for your answers.

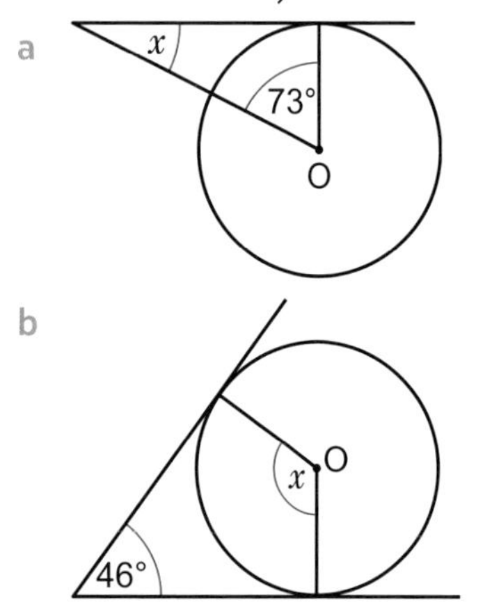

> **Q2 hint** 'Give a reason' means you need to state the geometric property you are using. For example
> Work out the value of x.
>
> $x = 180° - 90° - 54° = 36°$
> (angle in semicircle = 90°,
> angle sum of triangle is 180°)
>
> Look at the tangent properties in Key points 13 and 14.

3 Work out the sizes of angles x and y in each of these diagrams.
Give reasons for your answers.

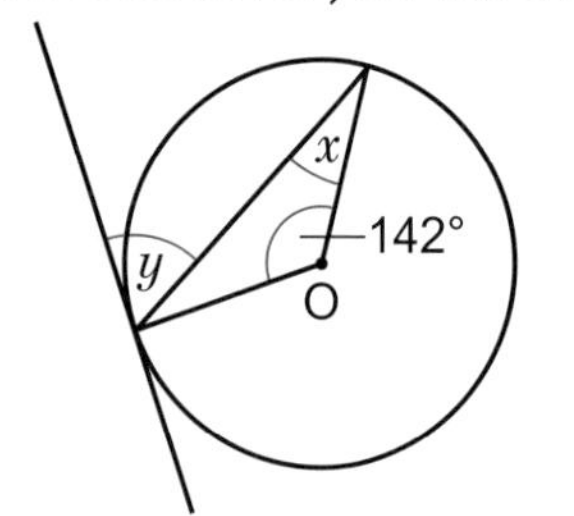

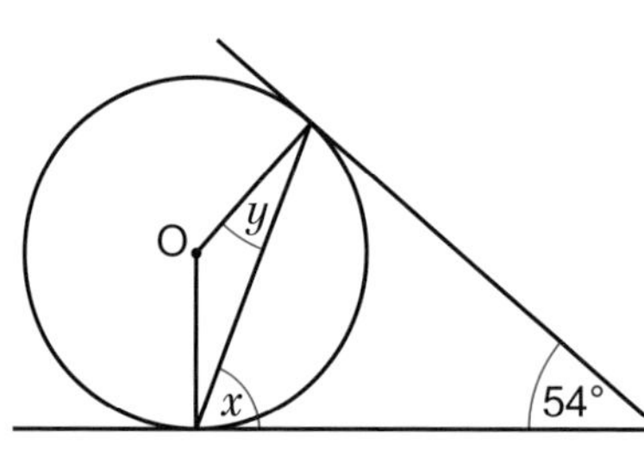

> **Q3** A common error is to not give reasons for the statements in a proof.

4 Work out the size of each angle marked with a letter.
Give a reason for each step of your working.

> **Q4 hint** If the centre (O) is marked, look for circle properties that involve the centre.

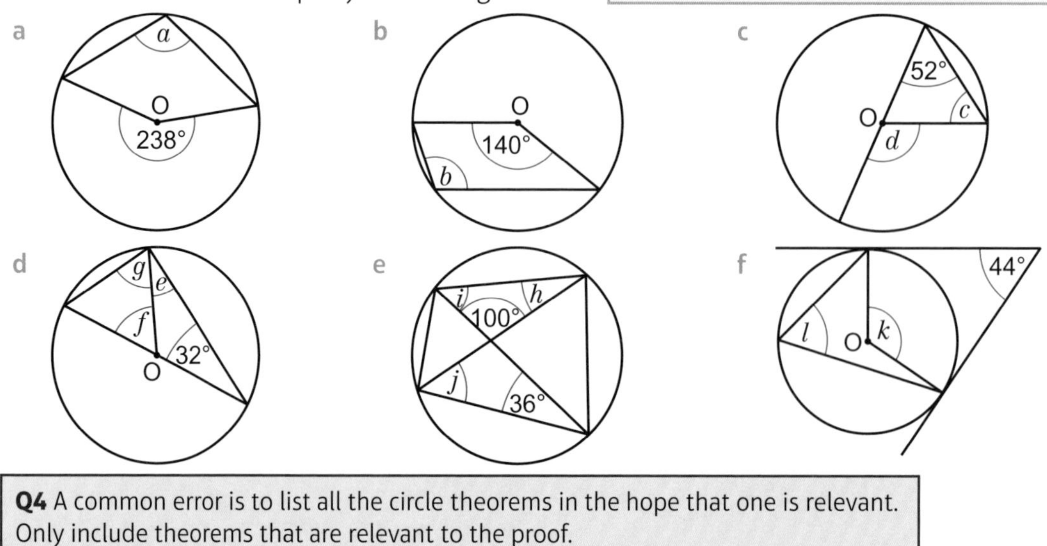

> **Q4** A common error is to list all the circle theorems in the hope that one is relevant.
> Only include theorems that are relevant to the proof.

5 **Reasoning** In the diagram, BOC is a diameter of the circle.

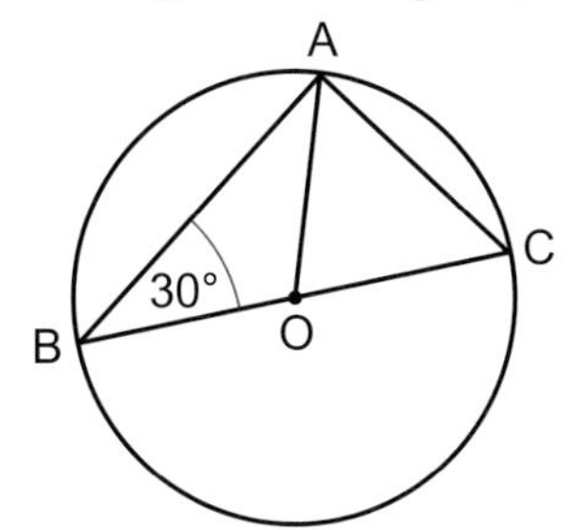

Q5 hint Remember that 'Show that …' means that you need to give reasons for every step of your working to show *how* the given result is obtained.

Show clearly that triangle OAC is equilateral.
Justify your answer.

6 Work out the size of each angle marked with a letter.
Give a reason for each step of your working.

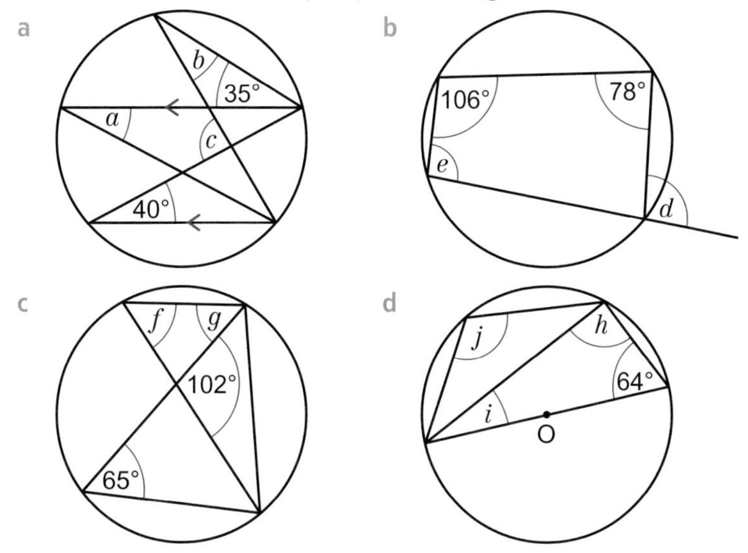

7 **Reasoning** In this diagram, points A, B, C and D lie on the circumference of a circle centre O.

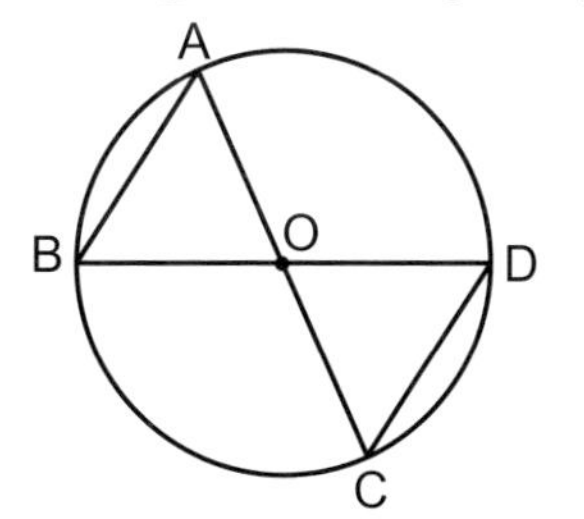

a Prove that angle BAO + angle CDO = angle BOC.

b Explain why all the angles at the edge of the circle are equal.

8 **Reasoning** In the diagram, ABCD is a cyclic quadrilateral.

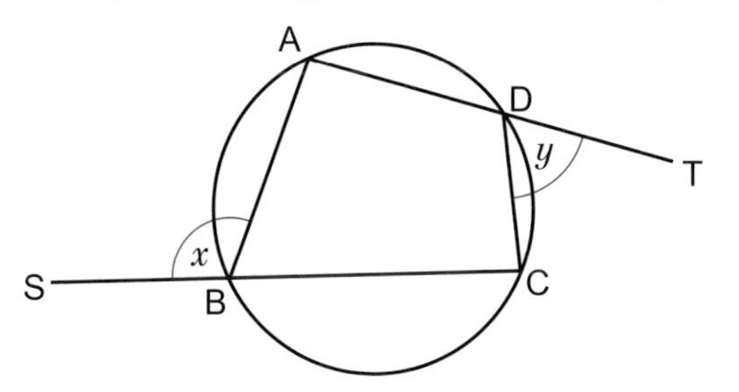

Prove that $x + y = 180°$.

9 Work out the size of each angle marked with a letter.
Give a reason for each step of your working.

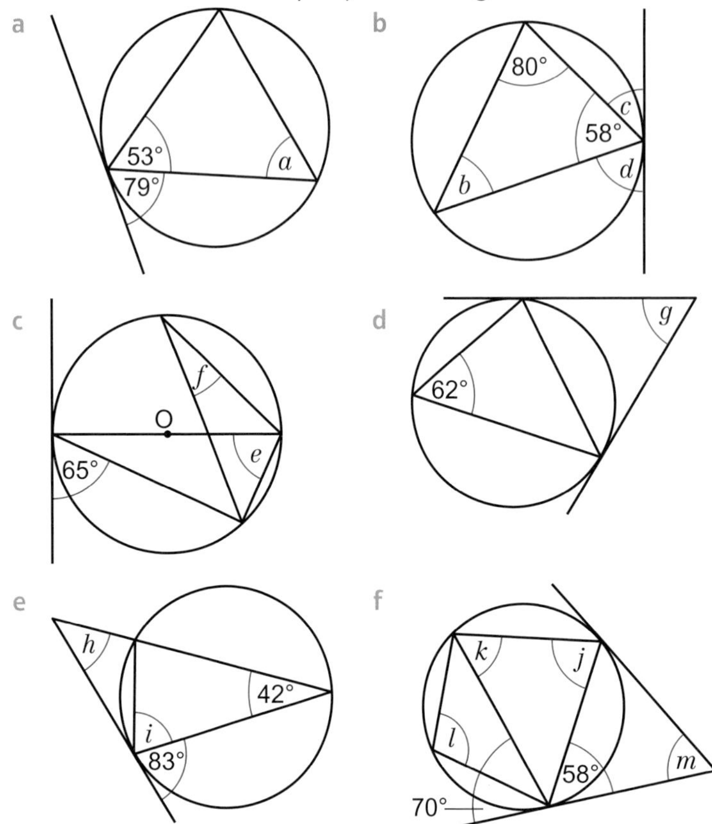

10 Problem-solving Points A, B, C and D lie on a circle with centre O.
CT is a tangent to the circle at C.
CA bisects angle BAD.

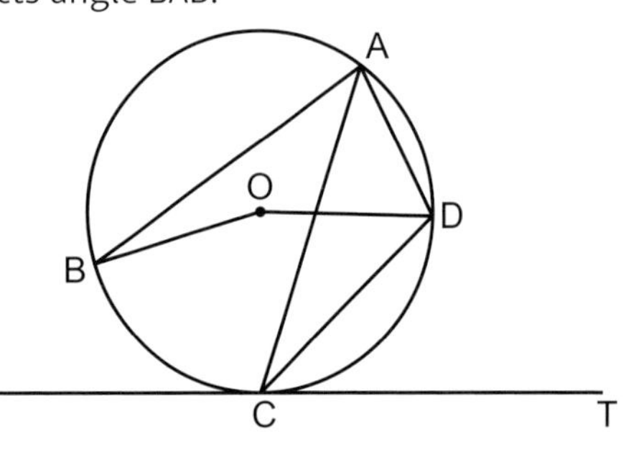

Prove that angle DCT is $\frac{1}{4}$ of angle BOD.

11 Problem-solving Work out the value of x in this diagram.

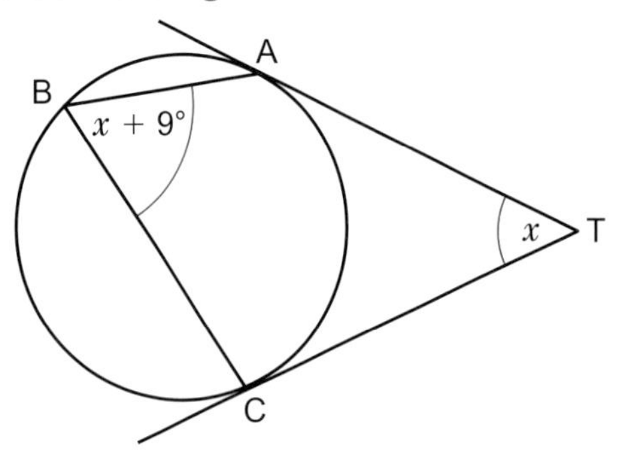

12 Problem-solving TBP and TCQ are tangents to the circle with centre O.
Point A lies on the circumference of the circle.

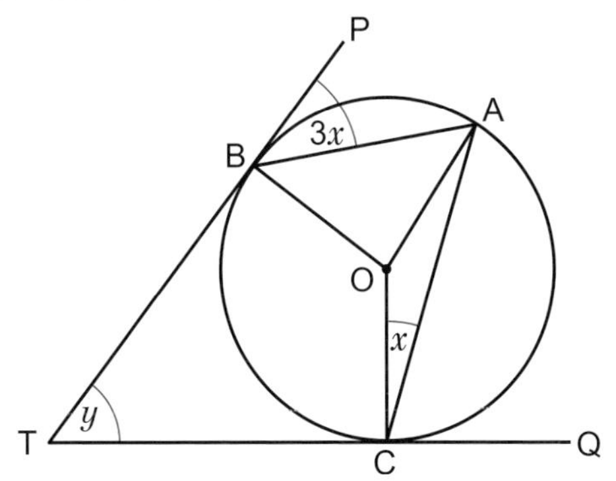

Prove that $y = 4x$.
Give reasons for any statements you make.

5.4 Trigonometry

Objectives

- Know the exact values of the trigonometric functions for angles of 0°, 30°, 45°, 60° and 90°.
- Use trigonometric formulae to solve problems in 2D and 3D, including finding the angle between a line and a plane.

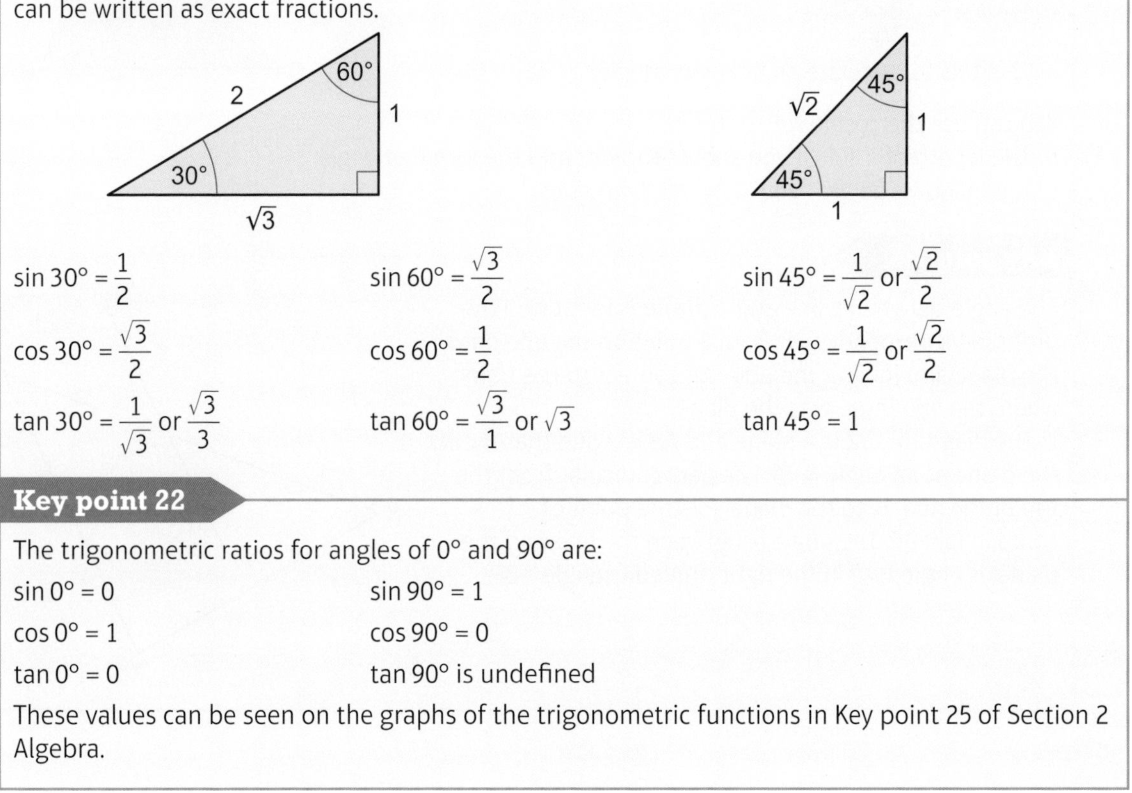

Key point 21

These right-angled triangles show how the values of sin, cos and tan for angles of 30°, 45° and 60° can be written as exact fractions.

$\sin 30° = \frac{1}{2}$ | $\sin 60° = \frac{\sqrt{3}}{2}$ | $\sin 45° = \frac{1}{\sqrt{2}}$ or $\frac{\sqrt{2}}{2}$

$\cos 30° = \frac{\sqrt{3}}{2}$ | $\cos 60° = \frac{1}{2}$ | $\cos 45° = \frac{1}{\sqrt{2}}$ or $\frac{\sqrt{2}}{2}$

$\tan 30° = \frac{1}{\sqrt{3}}$ or $\frac{\sqrt{3}}{3}$ | $\tan 60° = \frac{\sqrt{3}}{1}$ or $\sqrt{3}$ | $\tan 45° = 1$

Key point 22

The trigonometric ratios for angles of 0° and 90° are:

$\sin 0° = 0$ $\quad$ $\sin 90° = 1$

$\cos 0° = 1$ $\quad$ $\cos 90° = 0$

$\tan 0° = 0$ $\quad$ $\tan 90°$ is undefined

These values can be seen on the graphs of the trigonometric functions in Key point 25 of Section 2 Algebra.

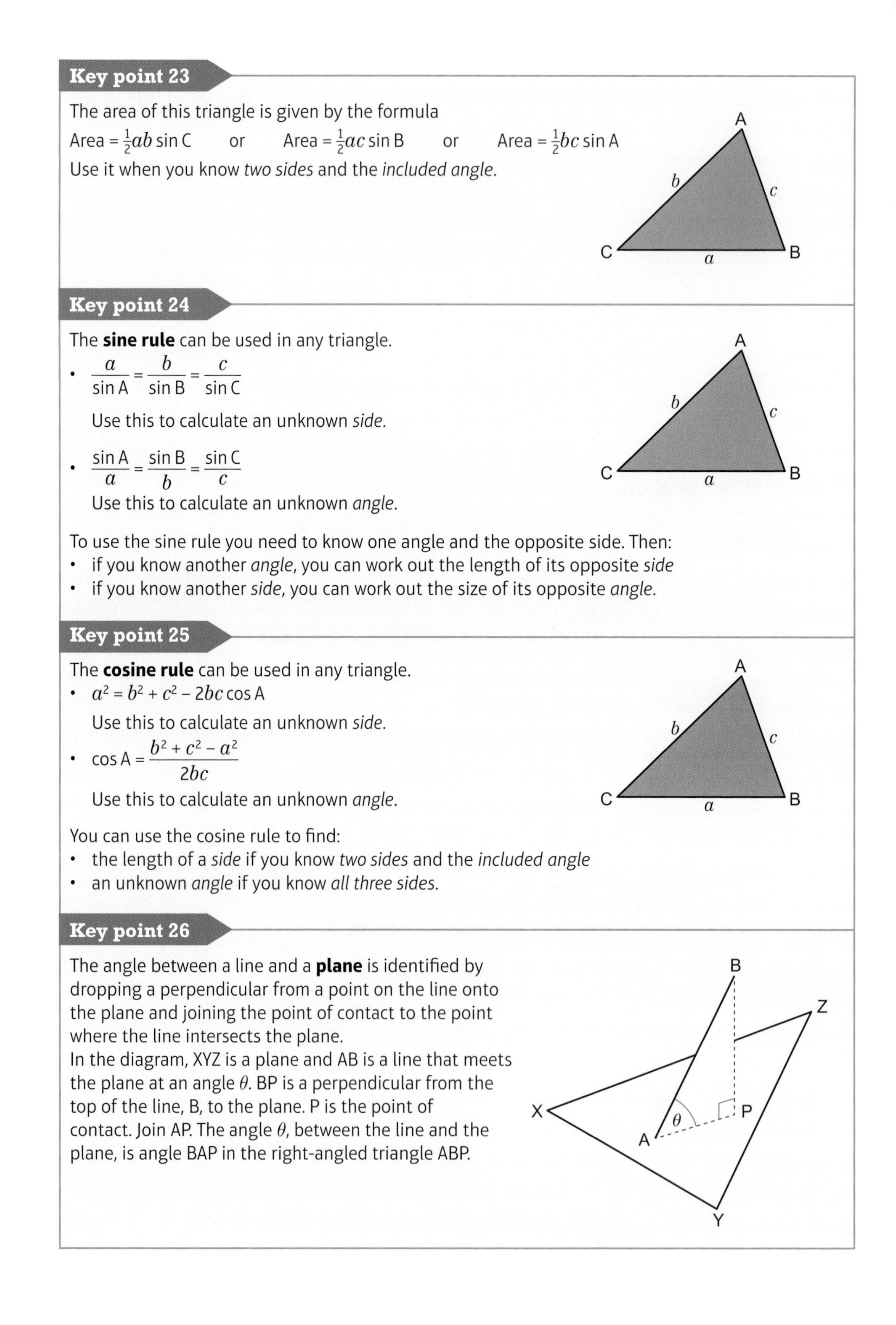

Key point 23

The area of this triangle is given by the formula

Area = $\frac{1}{2}ab \sin C$ or Area = $\frac{1}{2}ac \sin B$ or Area = $\frac{1}{2}bc \sin A$

Use it when you know *two sides* and the *included angle.*

Key point 24

The **sine rule** can be used in any triangle.

- $\frac{a}{\sin A} = \frac{b}{\sin B} = \frac{c}{\sin C}$

  Use this to calculate an unknown *side.*

- $\frac{\sin A}{a} = \frac{\sin B}{b} = \frac{\sin C}{c}$

  Use this to calculate an unknown *angle.*

To use the sine rule you need to know one angle and the opposite side. Then:

- if you know another *angle*, you can work out the length of its opposite *side*
- if you know another *side*, you can work out the size of its opposite *angle.*

Key point 25

The **cosine rule** can be used in any triangle.

- $a^2 = b^2 + c^2 - 2bc \cos A$

  Use this to calculate an unknown *side.*

- $\cos A = \frac{b^2 + c^2 - a^2}{2bc}$

  Use this to calculate an unknown *angle.*

You can use the cosine rule to find:

- the length of a *side* if you know *two sides* and the *included angle*
- an unknown *angle* if you know *all three sides.*

Key point 26

The angle between a line and a **plane** is identified by dropping a perpendicular from a point on the line onto the plane and joining the point of contact to the point where the line intersects the plane.

In the diagram, XYZ is a plane and AB is a line that meets the plane at an angle θ. BP is a perpendicular from the top of the line, B, to the plane. P is the point of contact. Join AP. The angle θ, between the line and the plane, is angle BAP in the right-angled triangle ABP.

1 Calculate the length of the side marked x in each of these triangles.
Give your answers correct to 3 s.f.

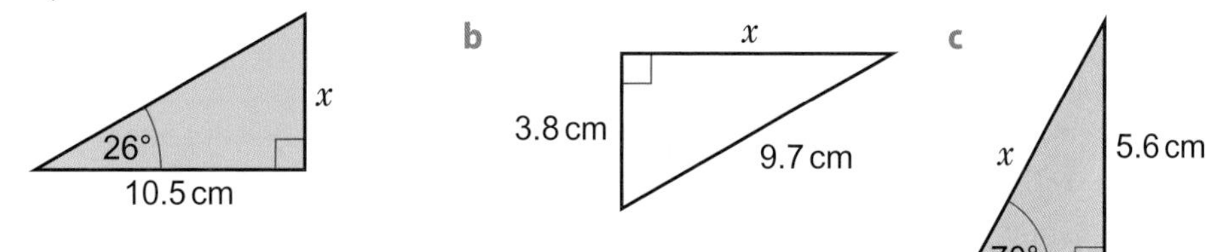

2 Calculate the size of the angle marked θ in each of these triangles.
Give your answers correct to 1 d.p.

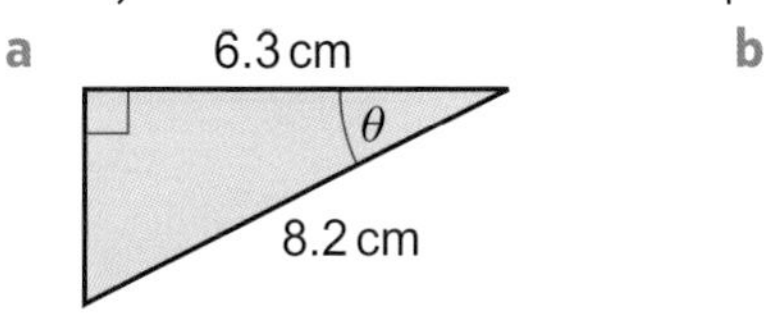

3 Work out the area of this sector.
Give your answer in terms of π.

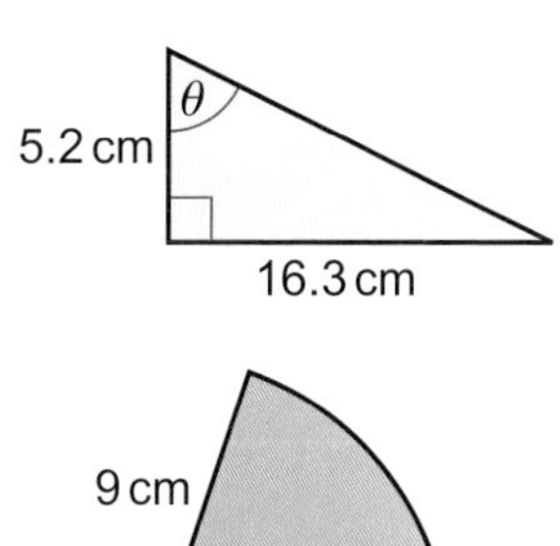

4 Calculate the area of each of these triangles.
Give your answers correct to 3 s.f.

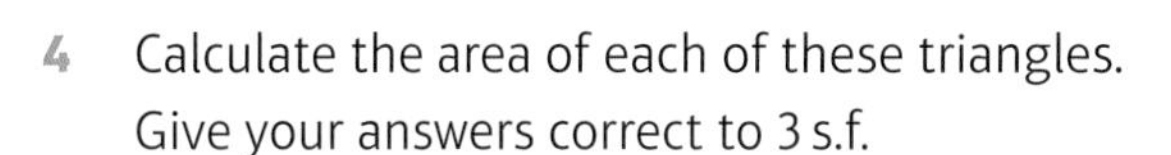

Q4 hint Use Area = $\frac{1}{2}ab\sin C$

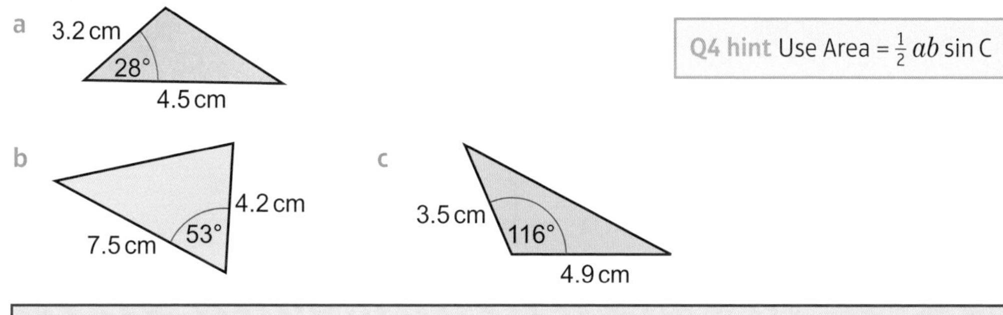

Q4 A common error is to have calculators set in radian or gradian mode instead of degree mode.

5 In each of these, you are given the area of the triangle.
Work out the length of the side marked x.
Give your answers correct to 3 s.f.

Q5 hint Use Area = $\frac{1}{2}ab\sin C$, but rearrange the formula to work out x.

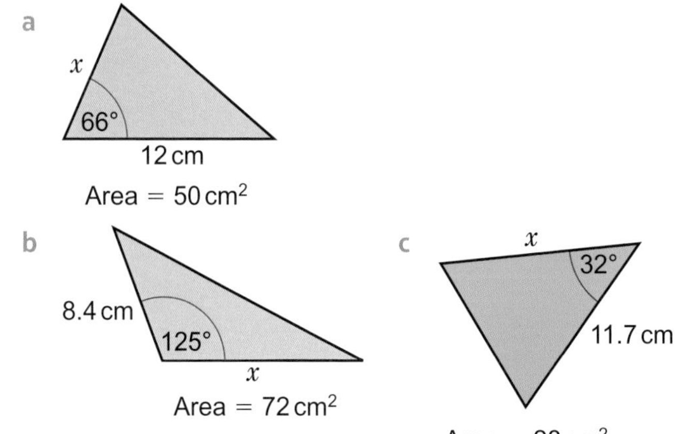

Warm up

6 Work out the area of the minor segment (shaded).
Give your answer correct to 3 s.f.

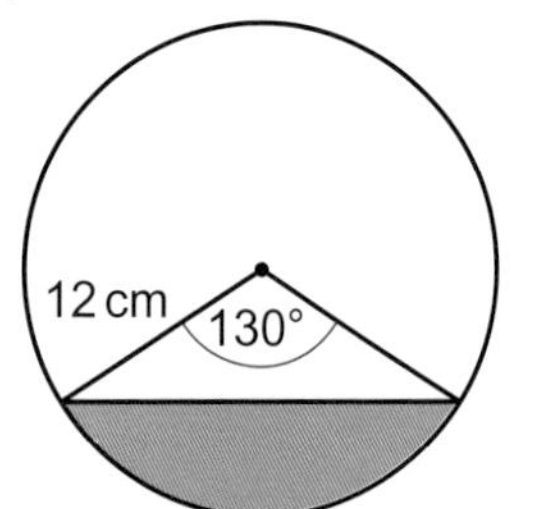

Q6 hint Area of segment = area of sector – area of triangle

7 Calculate the length of the side marked x in each of these diagrams.
Give your answers correct to 3 s.f.

Q7 hint Use the sine rule to calculate an unknown side.

a
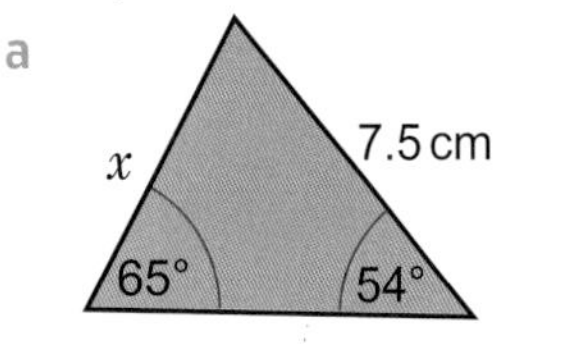

b
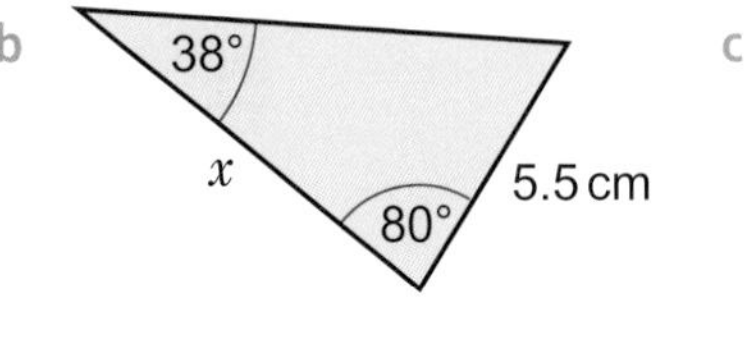

c
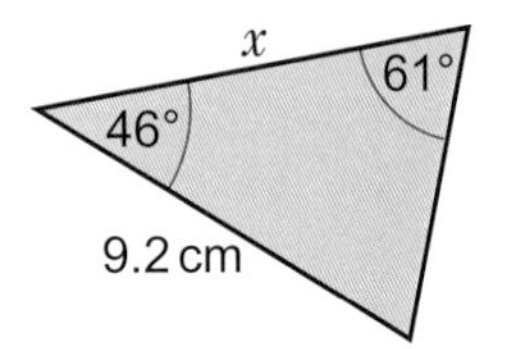

8 Calculate the size of the angle marked θ in each of these diagrams.
Give your answers correct to 3 s.f.

Q8 hint Use the sine rule to calculate an unknown angle.

a
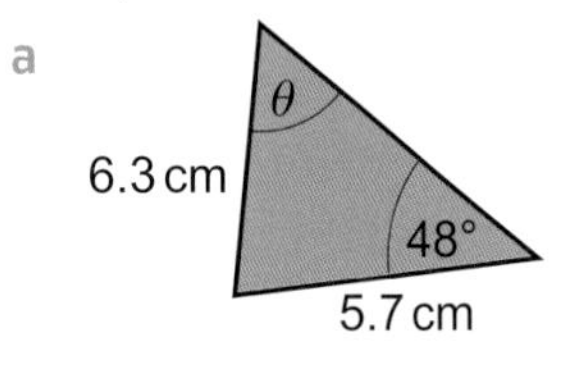

b
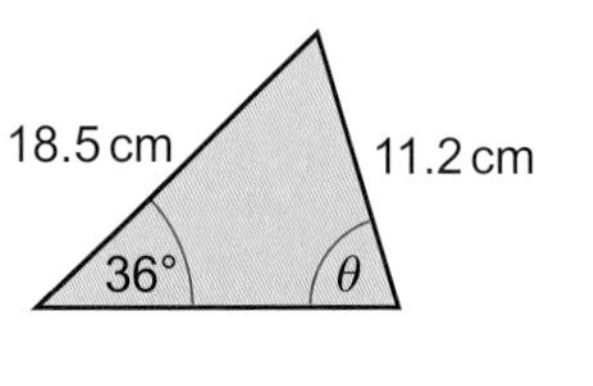

c
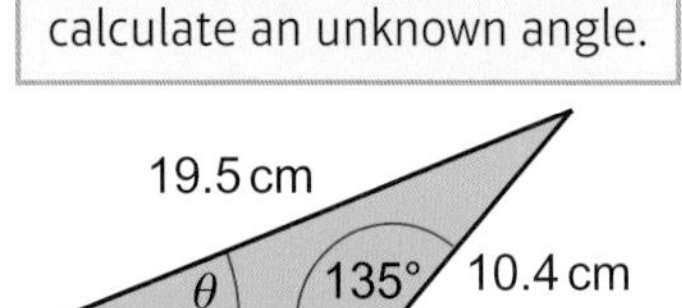

9 **Reasoning**
a Work out the length of AC.
b Work out the size of angle ACD, marked θ on the diagram.
c Work out the area of triangle ACD.
Give all your answers correct to 3 s.f.

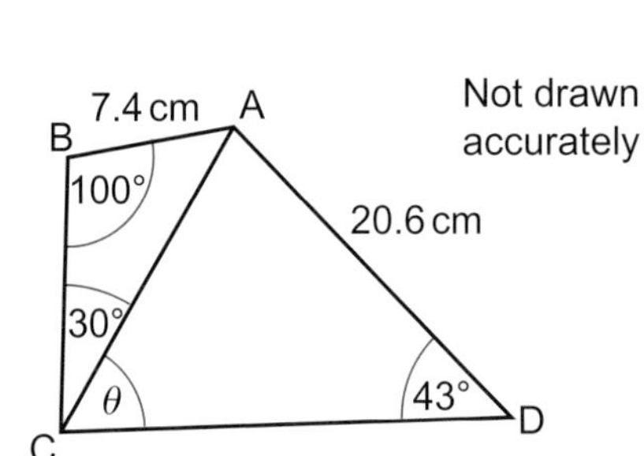

10 Calculate the length of the side marked x in each of these diagrams.
Give your answers correct to 3 s.f.

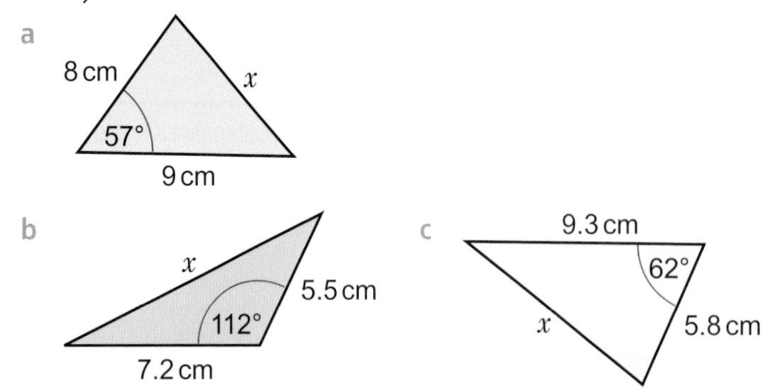

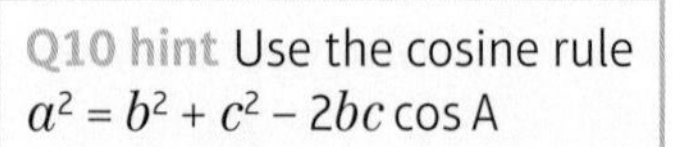
Q10 hint Use the cosine rule
$a^2 = b^2 + c^2 - 2bc \cos A$

11 Calculate the size of the angle marked θ in each of these diagrams.
Give your answers correct to 3 s.f.

Q11 hint Use the cosine rule $\cos A = \frac{b^2 + c^2 - a^2}{2bc}$

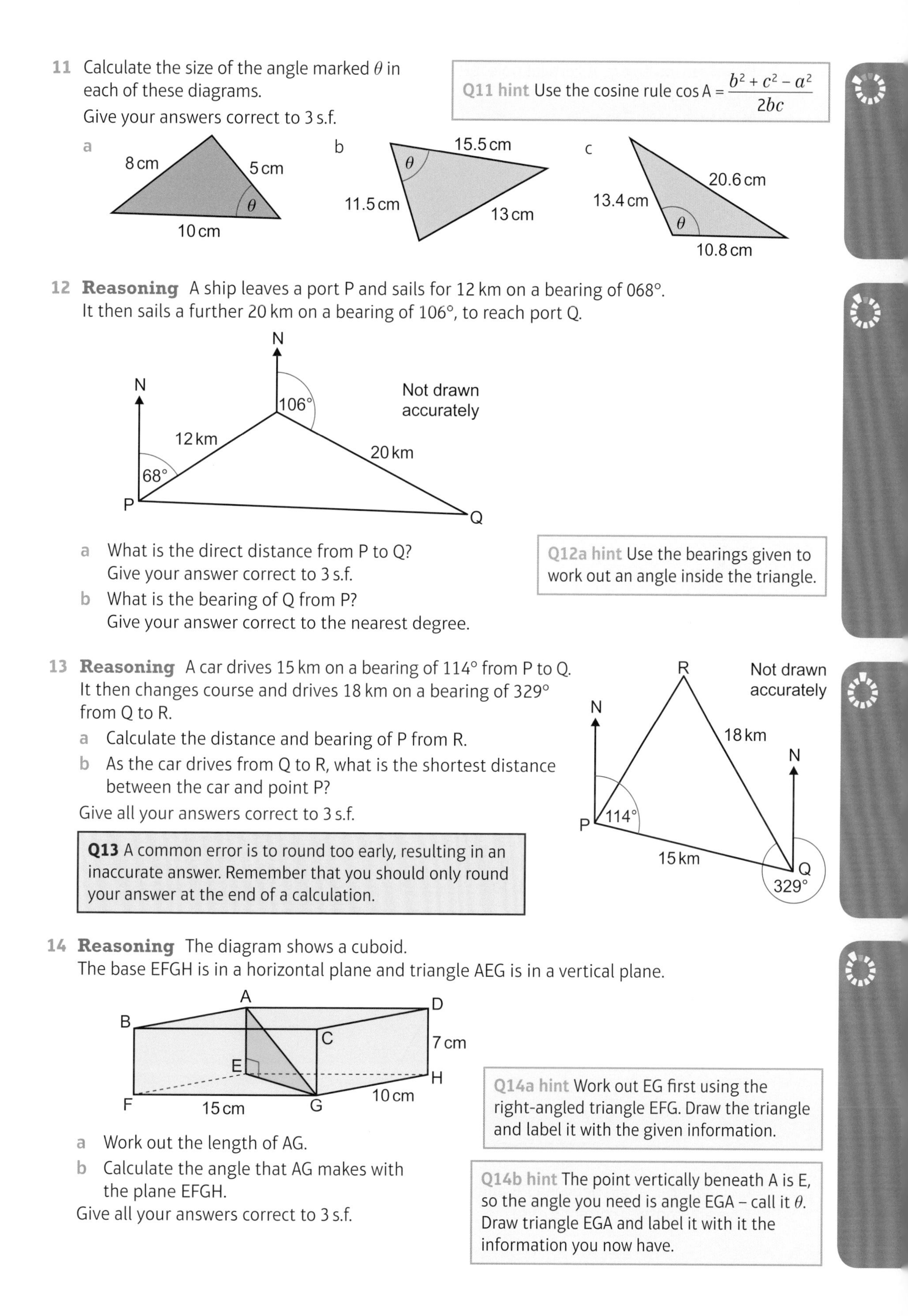

12 Reasoning A ship leaves a port P and sails for 12 km on a bearing of 068°. It then sails a further 20 km on a bearing of 106°, to reach port Q.

a What is the direct distance from P to Q?
Give your answer correct to 3 s.f.

b What is the bearing of Q from P?
Give your answer correct to the nearest degree.

Q12a hint Use the bearings given to work out an angle inside the triangle.

13 Reasoning A car drives 15 km on a bearing of 114° from P to Q. It then changes course and drives 18 km on a bearing of 329° from Q to R.

a Calculate the distance and bearing of P from R.

b As the car drives from Q to R, what is the shortest distance between the car and point P?

Give all your answers correct to 3 s.f.

Q13 A common error is to round too early, resulting in an inaccurate answer. Remember that you should only round your answer at the end of a calculation.

14 Reasoning The diagram shows a cuboid.
The base EFGH is in a horizontal plane and triangle AEG is in a vertical plane.

a Work out the length of AG.

b Calculate the angle that AG makes with the plane EFGH.

Give all your answers correct to 3 s.f.

Q14a hint Work out EG first using the right-angled triangle EFG. Draw the triangle and label it with the given information.

Q14b hint The point vertically beneath A is E, so the angle you need is angle EGA – call it θ. Draw triangle EGA and label it with it the information you now have.

15 **Problem-solving** A vertical pole, CP, stands at one corner of a rectangular, horizontal field.
AB = 40 m
AD = 30 m
Angle PDC = 25°

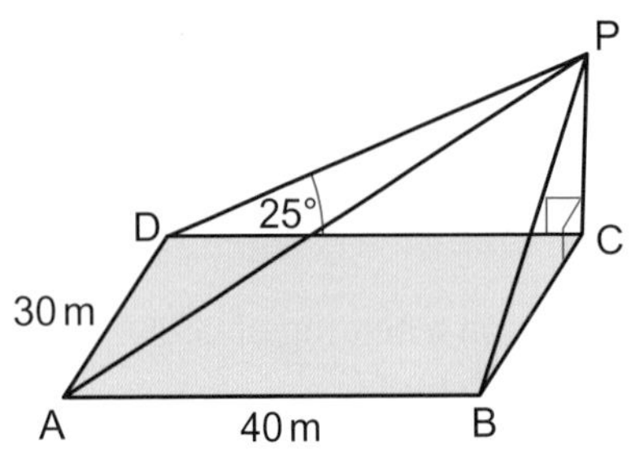

Calculate

a the height of the pole, CP
b the angle of elevation of P from B
c the angle of elevation of P from A.

Give all your answers correct to 3 s.f.

Q15b, c hint The angle of elevation of P from B is the angle between PB and the horizontal plane ABCD. Similarly for PA.

Q15c hint What length must be calculated first?

16 **Problem-solving** The diagram shows a rectangular-based pyramid.
The vertex, V, is vertically above M, the centre of the base, WXYZ.
The base lies in a horizontal plane.
WX = 32 cm, XY = 24 cm and VW = VX = VY = VZ = 27 cm.
N is the midpoint of XY.

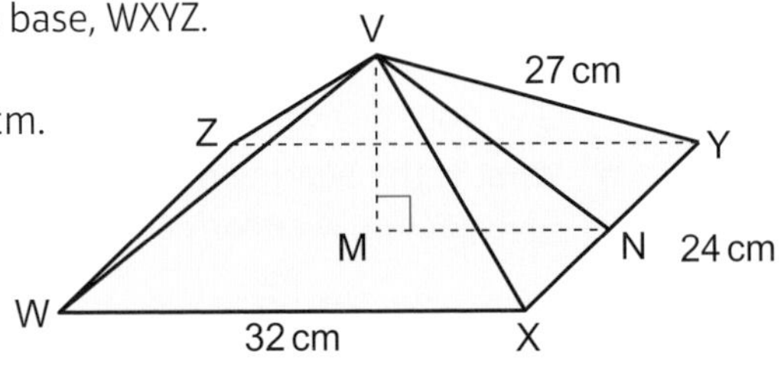

Calculate

a the length of WY
b the length of VM
c the angle that VY makes with the base WXYZ
d the length of VN
e the angle that VN makes with the base WXYZ.

Give all your answers correct to 3 s.f.

Q16 hint For each part, draw the right-angled triangle you are working on and label it with the information you have.

17 **Problem-solving** The diagram shows a wedge. AB = 14 cm and BC = 18 cm.
ABCD is a rectangle in a horizontal plane.
ADEF is a rectangle in a vertical plane.
P is one-third of the way from E to F.
PQ is perpendicular to AD.
The angle between PC and the plane ABCD is 30°.

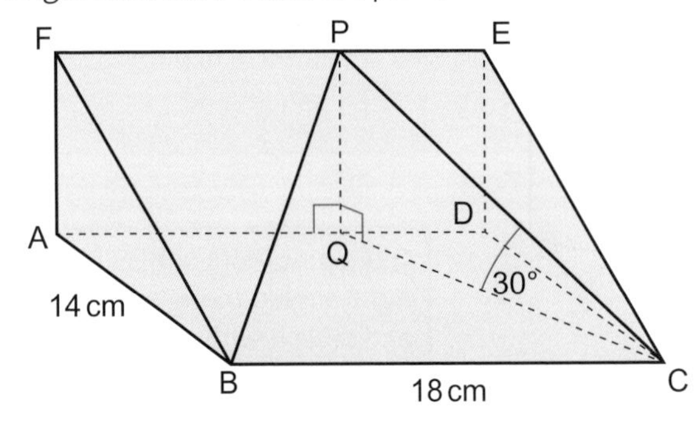

Calculate the angle between PB and the plane ABCD.
Give your answer correct to 3 s.f.

5.5 Vectors

Objectives

- Apply addition and subtraction of vectors, multiplication by a scalar, and diagrammatic and column representation of vectors to solve problems in 2D.
- Use vector methods to prove that lines are parallel or that points are collinear, and to construct geometric arguments and proofs.

Key point 27

A **vector** is a quantity that has magnitude and direction.

The **magnitude** of a vector is its size.

Displacement is change in position.

A displacement can be written as $\begin{pmatrix}3\\4\end{pmatrix}$ where 3 is the x-component and 4 is the y-component.

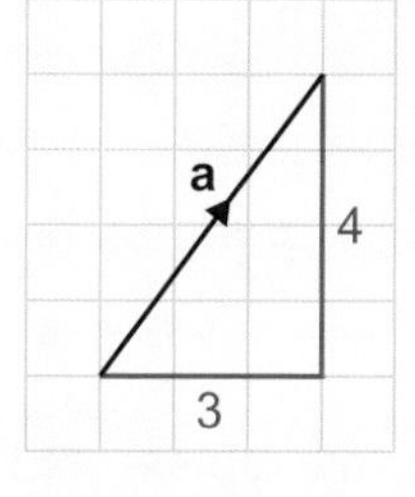

Examples of vectors are force (5 N acting vertically upwards) and velocity (15 km/h due north).

The displacement vector from A to B is written $\overrightarrow{AB}$.

Vectors are written as bold lower case letters, for example **a**, **b**, **c**

When handwriting, underline the letter, for example, $\underline{a}$, $\underline{b}$, $\underline{c}$

Key point 28

The magnitude of the vector $\begin{pmatrix}x\\y\end{pmatrix}$ is its length, i.e. $\sqrt{x^2 + y^2}$.

|a| means the magnitude of vector a. **|OA|** means the magnitude of vector $\overrightarrow{OA}$.

Equal vectors have the same magnitude and the same direction.

Key point 29

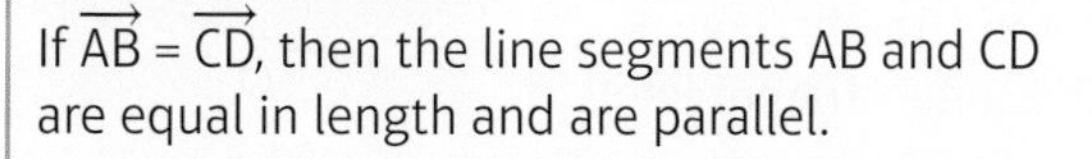

If $\overrightarrow{AB} = \overrightarrow{CD}$, then the line segments AB and CD are equal in length and are parallel.

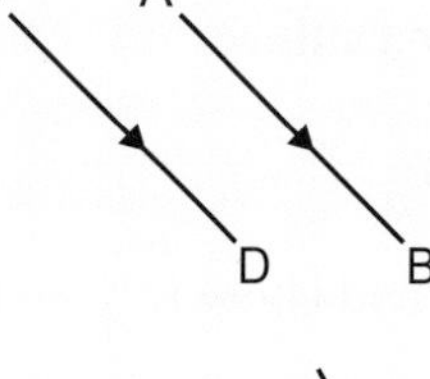

$\overrightarrow{AB} = -\overrightarrow{BA}$

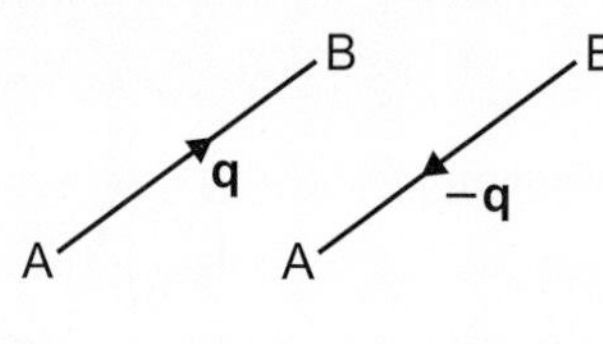

2**a** is twice as long as **a** and in the same direction.

−**a** is the same length as **a** but in the opposite direction.

a 2a −a

When a vector **a** is multiplied by a scalar k then the vector k**a** is parallel to **a** and is equal to k times **a**.

A **scalar** is a number, e.g. 3, 2, $\frac{1}{2}$, −1, …

Key point 30

The two-stage journey from A to B and then from B to C has the same starting point and the same finishing point as the single journey from A to C. So A to B followed by B to C is equivalent to A to C.

$\overrightarrow{AB} + \overrightarrow{BC} = \overrightarrow{AC}$

Triangle law for vector addition

Let $\overrightarrow{AB} = \mathbf{a}$, $\overrightarrow{BC} = \mathbf{b}$ and $\overrightarrow{AC} = \mathbf{c}$.
Then $\mathbf{a} + \mathbf{b} = \mathbf{c}$ forms a triangle.
$\mathbf{c}$ is called the **resultant vector** of the two vectors $\mathbf{a}$ and $\mathbf{b}$.

Key point 31

An alternative to the triangle law is the **parallelogram law for vector addition**.
In this parallelogram, $\overrightarrow{PQ} = \mathbf{a}$ and $\overrightarrow{PS} = \mathbf{b}$.
$\overrightarrow{SR} = \overrightarrow{PQ} = \mathbf{a}$ and $\overrightarrow{QR} = \overrightarrow{PS} = \mathbf{b}$

Using vector addition:

$\overrightarrow{PR} = \overrightarrow{PQ} + \overrightarrow{QR} = \mathbf{a} + \mathbf{b}$

$\overrightarrow{SQ} = \overrightarrow{SR} + \overrightarrow{RQ} = \mathbf{a} + -\mathbf{b} = \mathbf{a} - \mathbf{b}$

$\overrightarrow{QS} = \overrightarrow{QR} + \overrightarrow{RS} = \mathbf{b} + -\mathbf{a} = \mathbf{b} - \mathbf{a}$

The diagonals of the parallelogram are represented by the vectors $\mathbf{a} + \mathbf{b}$ and either $\mathbf{a} - \mathbf{b}$ or $\mathbf{b} - \mathbf{a}$ depending on the direction.

Key point 32

With the origin O, the vectors $\overrightarrow{OA}$ and $\overrightarrow{OB}$ are called **position vectors** of the points A and B.
In general, a point with coordinates (p, q) has position vector $\begin{pmatrix} p \\ q \end{pmatrix}$.

When $\overrightarrow{OA} = \mathbf{a}$ and $\overrightarrow{OB} = \mathbf{b}$, $\overrightarrow{AB} = \overrightarrow{AO} + \overrightarrow{OB} = \mathbf{b} - \mathbf{a}$.

Key point 33

$\overrightarrow{PQ} = k\overrightarrow{QR}$ shows that the lines PQ and QR are parallel. Both lines pass through point Q so PQ and QR are part of the same straight line.
Points P, Q and R are **collinear** (they all lie on the same straight line).

Key point 34

Vectors such as $\begin{pmatrix} a \\ b \end{pmatrix}$ and either $\begin{pmatrix} -b \\ a \end{pmatrix}$ or $\begin{pmatrix} b \\ -a \end{pmatrix}$ are perpendicular.

Warm up

1 Work out the lengths of the sides marked x and y.
Give your answers in surd form.

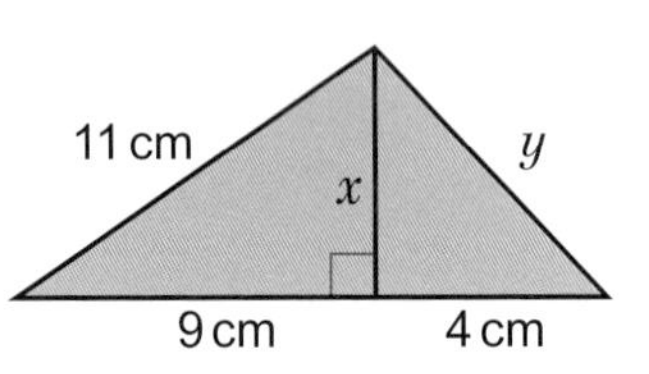

2 The vector $\begin{pmatrix} -7 \\ 12 \end{pmatrix}$ transforms shape A to shape B.

The vector $\begin{pmatrix} -3 \\ -5 \end{pmatrix}$ transforms shape B to shape C.

What vector transforms shape C to shape A?

> **Q2** A common error is to use the wrong notation for column vectors. Always write them like this $\begin{pmatrix} p \\ q \end{pmatrix}$, never as $\left(\frac{p}{q}\right)$ or (p, q).

3 Work out the magnitude of each of these vectors, leaving your answers in surd form.

a $\mathbf{a} = \begin{pmatrix} -5 \\ 9 \end{pmatrix}$ b $\mathbf{b} = \begin{pmatrix} 7 \\ -2 \end{pmatrix}$ c $\mathbf{c} = \begin{pmatrix} 12 \\ 3 \end{pmatrix}$ d $\mathbf{d} = \begin{pmatrix} -6 \\ -13 \end{pmatrix}$

4 $\mathbf{u} = \begin{pmatrix} 4 \\ -1 \end{pmatrix}$ $\mathbf{v} = \begin{pmatrix} -2 \\ 3 \end{pmatrix}$ $\mathbf{w} = \begin{pmatrix} 2 \\ 5 \end{pmatrix}$

On a square grid, draw diagrams to illustrate each of these vectors.

a $\mathbf{u} + \mathbf{w}$ b $\mathbf{u} - \mathbf{w}$ c $2\mathbf{v} + \mathbf{w}$

d $\mathbf{w} - 3\mathbf{u}$ e $\mathbf{u} + \mathbf{v} - 2\mathbf{w}$ f $\mathbf{w} - 2\mathbf{v} - 3\mathbf{u}$

5 $\mathbf{p} = \begin{pmatrix} 2 \\ -2 \end{pmatrix}$ $\mathbf{q} = \begin{pmatrix} 6 \\ 2 \end{pmatrix}$ $\mathbf{r} = \begin{pmatrix} -1 \\ 5 \end{pmatrix}$

$\mathbf{p} + 3\mathbf{r} = 2\mathbf{q} - \mathbf{s}$

a Work out **s** as a column vector.

b Calculated the magnitude of **s**, leaving your answer as a surd in the form $m\sqrt{n}$, where m and n are integers.

6 On this grid of congruent parallelograms, the origin is labelled O, $\overrightarrow{OA} = \mathbf{a}$ and $\overrightarrow{OB} = \mathbf{b}$.

C D E F G
K B J I H
L O A M N
T S R Q P
U V W X Y

> **Q6** A common error is to forget that a change in vector direction requires a change of signs.

Write, in terms of **a** and/or **b**

a $\overrightarrow{AB}$ b $\overrightarrow{OR}$ c $\overrightarrow{OV}$ d $\overrightarrow{OP}$

e $\overrightarrow{OG}$ f $\overrightarrow{PE}$ g $\overrightarrow{YT}$ h $\overrightarrow{GU}$

7 On this grid of congruent parallelograms, the origin is labelled O, $\overrightarrow{OA} = \mathbf{a}$ and $\overrightarrow{OB} = \mathbf{b}$.

Copy the grid. Mark points and write letters on the grid to represent each of these vectors.

a $\overrightarrow{OC} = 2\mathbf{b} - \mathbf{a}$ b $\overrightarrow{OD} = 2\mathbf{a} + 3\mathbf{b}$ c $\overrightarrow{OE} = \mathbf{a} - 3\mathbf{b}$ d $\overrightarrow{OF} = -2\mathbf{a} + \frac{1}{2}\mathbf{b}$

e $\overrightarrow{OG} = \mathbf{a} + \frac{3}{2}\mathbf{b}$ f $\overrightarrow{OH} = 2\mathbf{a} - \frac{1}{2}\mathbf{b}$ g $\overrightarrow{OJ} = \frac{5}{2}\mathbf{a} - 2\mathbf{b}$ h $\overrightarrow{OK} = -\frac{1}{2}\mathbf{a} - \frac{3}{2}\mathbf{b}$

8 **Reasoning** The points J, K, L and M have coordinates (–3, –1), (–6, –8), (14, 0) and (12, 5), respectively. O is the origin.

a Write down the position vectors $\overrightarrow{OJ}$ and $\overrightarrow{OM}$.

b Work out, as column vectors, $\overrightarrow{JM}$ and $\overrightarrow{KL}$.

c What do these results tell you about the lines JM and KL?

d Work out the column vector $\overrightarrow{LM}$.
What does this tell you about the relationship between $\overrightarrow{LM}$ and both $\overrightarrow{JM}$ and $\overrightarrow{KL}$?

e Fully describe quadrilateral JKLM.

9 **Reasoning** In triangle XYZ, $\overrightarrow{XY} = 2\mathbf{a}$ and $\overrightarrow{XZ} = 3\mathbf{b}$.
S is the midpoint of YZ.
T is the point on XZ such that XT : TZ = 1 : 2

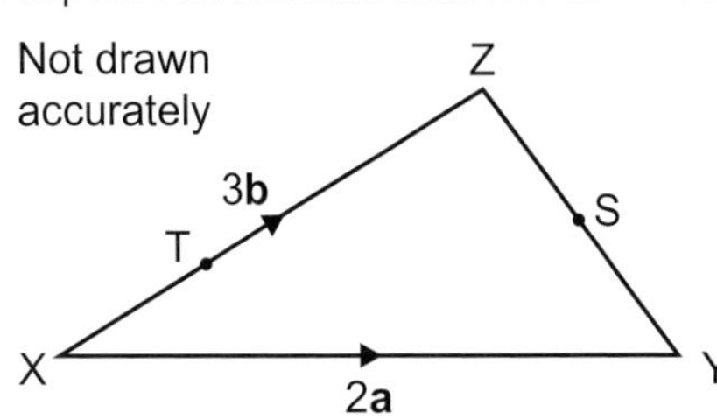

Write each of these vectors in terms of **a** and/or **b**.
Simplify your answers.

a $\overrightarrow{YZ}$ b $\overrightarrow{YS}$ c $\overrightarrow{XS}$

d $\overrightarrow{XT}$ e $\overrightarrow{YT}$ f $\overrightarrow{TS}$

Q9c hint Make use of the vectors you already know. $\overrightarrow{XS} = \overrightarrow{XY} + \overrightarrow{YS}$. Simplify your answer fully.

Q9d A common error is to misunderstand ratios.
XT : TZ = 1 : 2, so XT is $\frac{1}{3}$ of XZ, not $\frac{1}{2}$.

Q9f hint Use the 'nose to tail' rule when doing vector addition, where the last letter of one vector is the first letter of the next.
$\overrightarrow{TS} = \overrightarrow{TZ} + \overrightarrow{ZS}$
or $\overrightarrow{TS} = \overrightarrow{TX} + \overrightarrow{XS}$
or $\overrightarrow{TS} = \overrightarrow{TX} + \overrightarrow{XY} + \overrightarrow{YS}$.
You could use any of these.

10 **Reasoning** PQRS is a quadrilateral.
M is the midpoint of QR. N is the midpoint of RS.
$\overrightarrow{PQ} = \mathbf{a}$, $\overrightarrow{PR} = \mathbf{b}$ and $\overrightarrow{PS} = \mathbf{c}$

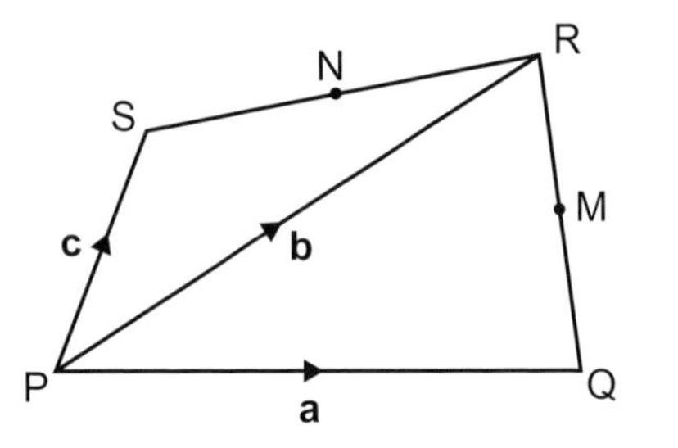

Q10a hint Use the triangle law of vector addition.
$\overrightarrow{QR} = \overrightarrow{QP} + \overrightarrow{PR}$

Q10g hint $\overrightarrow{PN} = \overrightarrow{PS} + \overrightarrow{SN} = \overrightarrow{PS} + \frac{1}{2}\overrightarrow{SR}$

Write each of these vectors in terms of **a**, **b** and/or **c**. Simplify your answers.

a $\overrightarrow{QR}$ b $\overrightarrow{RQ}$ c $\overrightarrow{SR}$ d $\overrightarrow{SQ}$

e $\overrightarrow{QM}$ f $\overrightarrow{PM}$ g $\overrightarrow{PN}$ h $\overrightarrow{NM}$

11 **Problem-solving** WXYZ is a trapezium, with XY parallel to WZ.
$\overrightarrow{XZ} = 5\mathbf{a} - 2\mathbf{b}$, $\overrightarrow{ZY} = 3\mathbf{a} + 4\mathbf{b}$ and $\overrightarrow{WZ} = 6\mathbf{a} + k\mathbf{b}$, where k is a number to be determined.

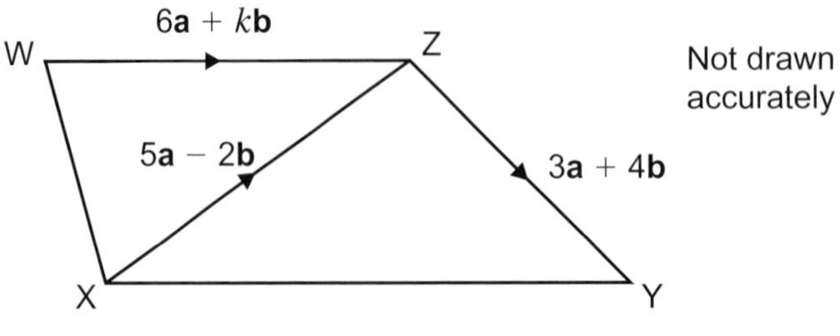

Work out the value of k.
Show working to justify your answer.

12 Problem-solving OACB is a trapezium in which $\overrightarrow{OA} = 3\mathbf{a}$, $\overrightarrow{OB} = 2\mathbf{b}$ and $\overrightarrow{AC} = 4\mathbf{b}$.

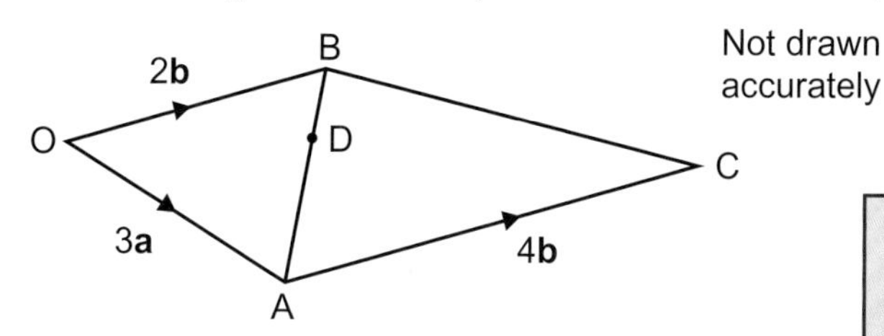

Q12 A common error when proving points are collinear (are on a straight line) is to forget you need to show that one vector is a multiple of the other *and* that they share a common point.

a Write in terms of **a** and **b**

i $\overrightarrow{OC}$ ii $\overrightarrow{AB}$

b D lies on BA such that BD : DA = 1 : 2.

Write $\overrightarrow{OD}$ in terms of **a** and **b**, giving your answer in its simplest form.

c Show clearly that ODC is a straight line.

13 Problem-solving OAB is a triangle.

M is the midpoint of OA. P and Q are points of trisection of AB.

$\overrightarrow{OA} = 3\mathbf{a}$ and $\overrightarrow{OB} = 3\mathbf{b}$

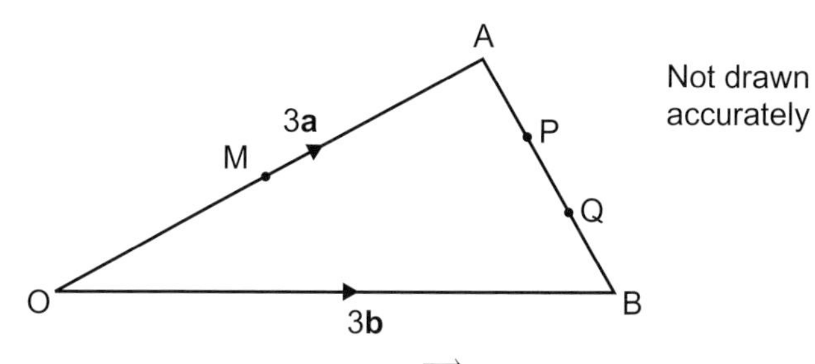

a Write down an expression for $\overrightarrow{AB}$ in terms of **a** and **b**.

b Show clearly that $\overrightarrow{MP} = \frac{1}{2}\mathbf{a} + \mathbf{b}$.

c Write down, and simplify, an expression for $\overrightarrow{OQ}$ in terms of **a** and **b**.

d What do your answers tell you about quadrilateral OMPQ?

14 Problem-solving The diagram shows a square OAPB.

$\overrightarrow{OA} = \mathbf{a}$ and $\overrightarrow{OB} = \mathbf{b}$. M is the midpoint of BP.

N is the point on AM such that AN : NM = 1 : 2

BP is extended to Q where BQ $= 2\frac{1}{2}$BP.

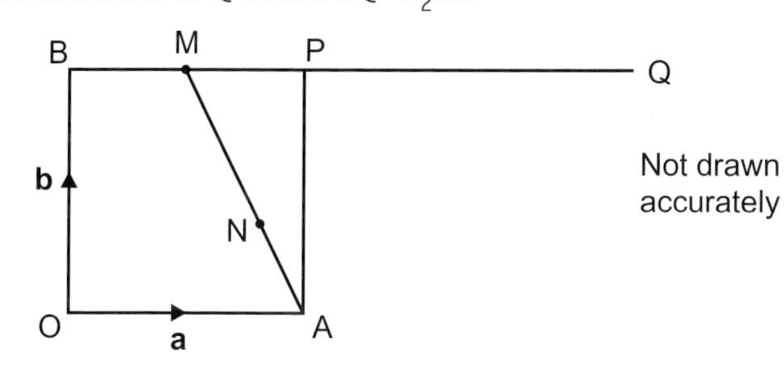

a Write these vectors in terms of **a** and **b**.

Give your answers in their simplest form.

i $\overrightarrow{OQ}$ ii $\overrightarrow{AM}$ iii $\overrightarrow{AN}$ iv $\overrightarrow{ON}$

b What can you deduce about points O, N and Q?

Show working to justify your answer.

5.6 Mixed exercise

Objective

- Consolidate your learning with more practice.

1 **Reasoning** A, B, C and D are points on the circumference of a circle with centre O.
Work out the sizes of the angles marked x and y.
Give reasons for any statements you make.

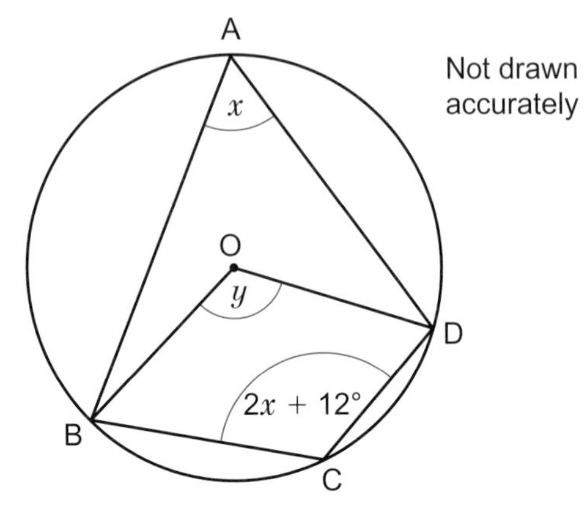

2 Draw a coordinate grid from −8 to 8 on both axes.
Draw trapezium A at (1, 2), (4, 2), (4, 4) and (3, 4).

a Rotate A 90° anticlockwise about (−1, 3). Label the image B.

b Enlarge B by scale factor −2, centre of enlargement (−2, 4). Label the image C.

c Translate C by the vector $\begin{pmatrix} 5 \\ -4 \end{pmatrix}$. Label the image D.

d Reflect D in the line $y = -x$. Label the image E.

e Enlarge E by scale factor $\frac{1}{2}$, centre of enlargement (−2, −1). Label the image F.

f Rotate F 90° clockwise about (−5, 0). Label the image G.

g Reflect G in the line $y = -1$. Label the image H.

h Translate H by the vector $\begin{pmatrix} 13 \\ -2 \end{pmatrix}$. Label the image J.

i Describe the single transformation that maps J onto A.

3 **Exam question**

The diagram shows a swimming pool in the shape of a prism.

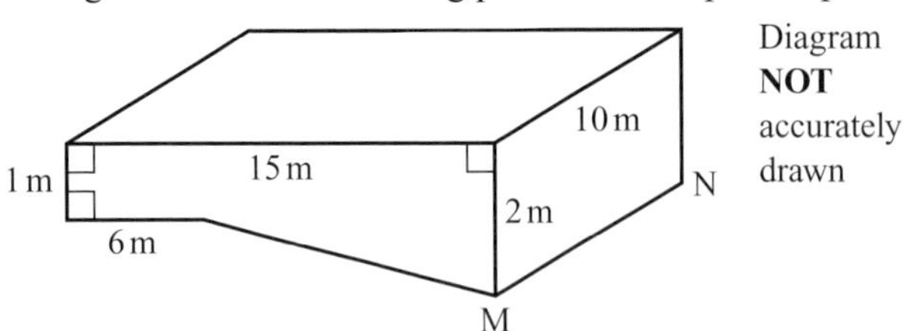

The swimming pool is empty.

The swimming pool is filled with water at a constant rate of 50 litres per minute.

a Work out how long it will take for the swimming pool to be completely full of water.

Give your answer in hours.

($1\text{ m}^3 = 1000$ litres) **(5 marks)**

November 2014, Q13a, 1MA0/2H

Exam hint

Many students did not divide the cross-section face correctly. Take care when dealing with compound shapes to calculate, and use the correct dimensions.

4 **Reasoning** In the diagram, PAQ is the tangent at A to a circle with centre O.
Angle BOC = 112°
Angle CAQ = 73°

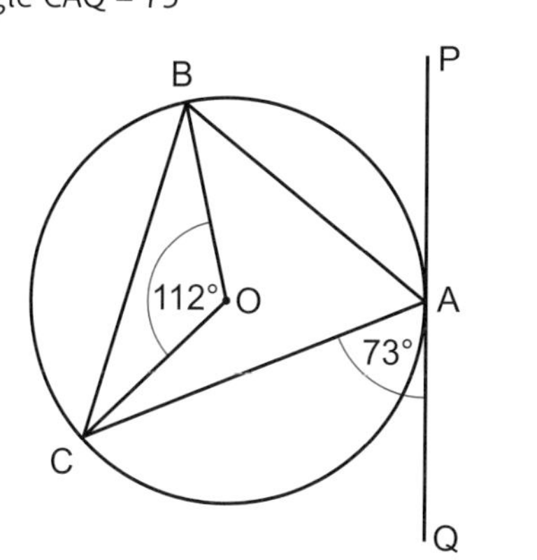

Work out the size of angle OBA.
Show your working, giving reasons for any statements you make.

5 **Exam question**

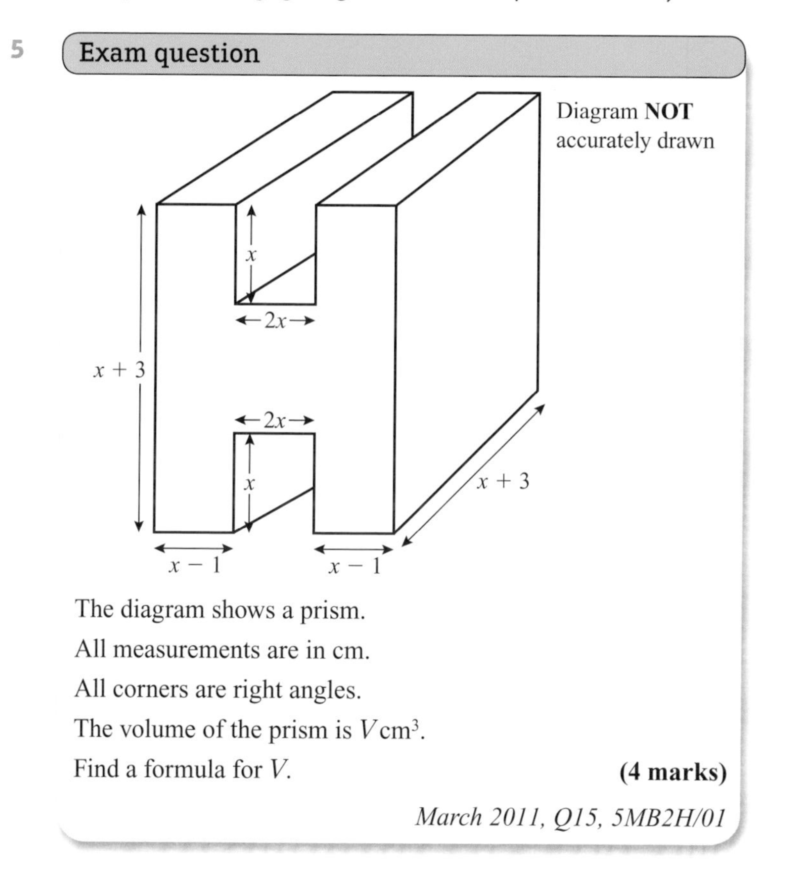

The diagram shows a prism.
All measurements are in cm.
All corners are right angles.
The volume of the prism is V cm^3.
Find a formula for V. **(4 marks)**

March 2011, Q15, 5MB2H/01

Exam hint
Many students failed to use brackets correctly when expanding expressions. Brackets help with organisation of the separate areas.

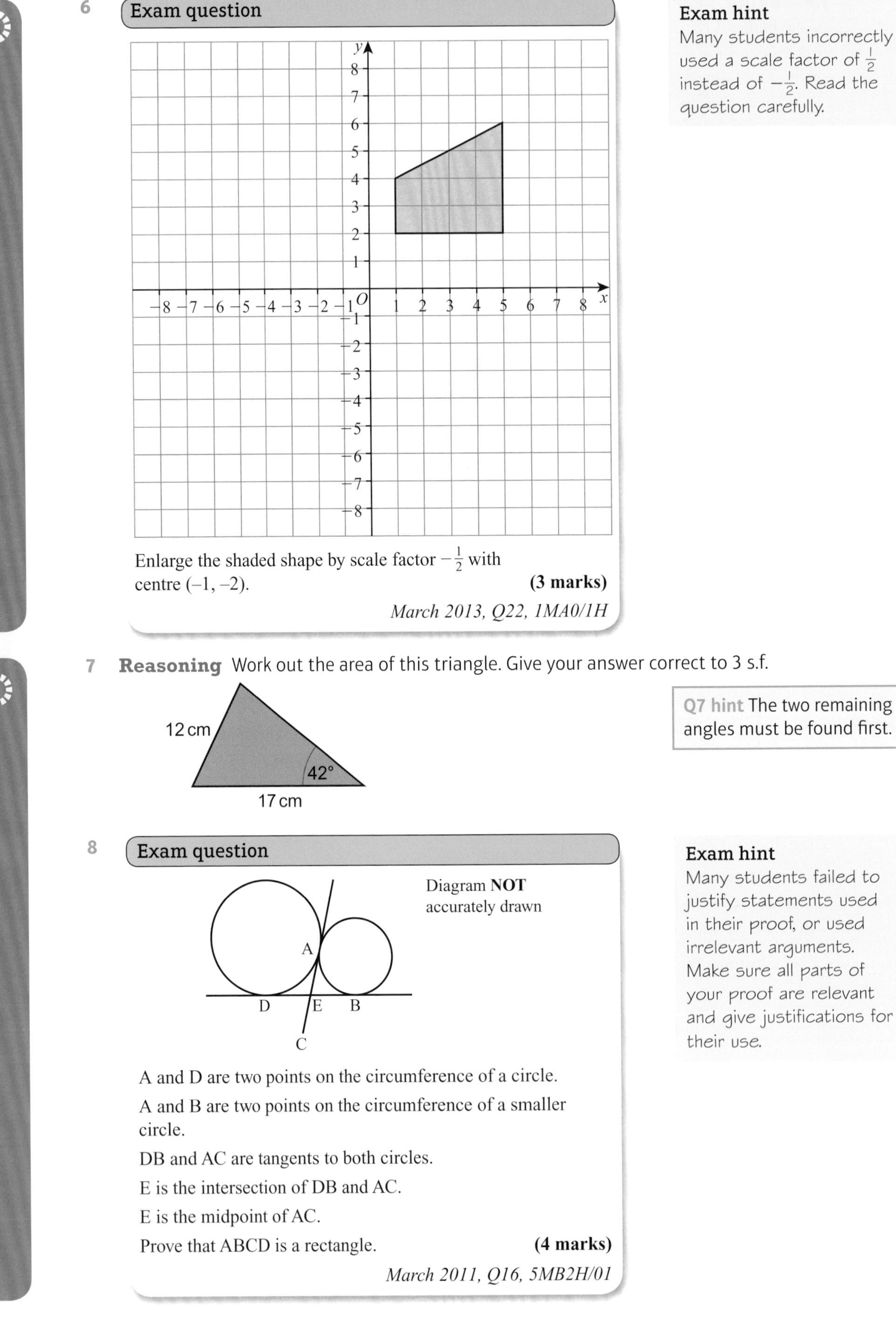

6

Exam question

Enlarge the shaded shape by scale factor $-\frac{1}{2}$ with centre (–1, –2). **(3 marks)**

March 2013, Q22, 1MA0/1H

Exam hint

Many students incorrectly used a scale factor of $\frac{1}{2}$ instead of $-\frac{1}{2}$. Read the question carefully.

7 **Reasoning** Work out the area of this triangle. Give your answer correct to 3 s.f.

Q7 hint The two remaining angles must be found first.

8

Exam question

A and D are two points on the circumference of a circle.

A and B are two points on the circumference of a smaller circle.

DB and AC are tangents to both circles.

E is the intersection of DB and AC.

E is the midpoint of AC.

Prove that ABCD is a rectangle. **(4 marks)**

March 2011, Q16, 5MB2H/01

Exam hint

Many students failed to justify statements used in their proof, or used irrelevant arguments. Make sure all parts of your proof are relevant and give justifications for their use.

9 Reasoning The diagram shows a cyclic quadrilateral ABDE.

BC = BE

Angle EAB = $3x$

Angle EDB = $9x$

Angle DBE = x

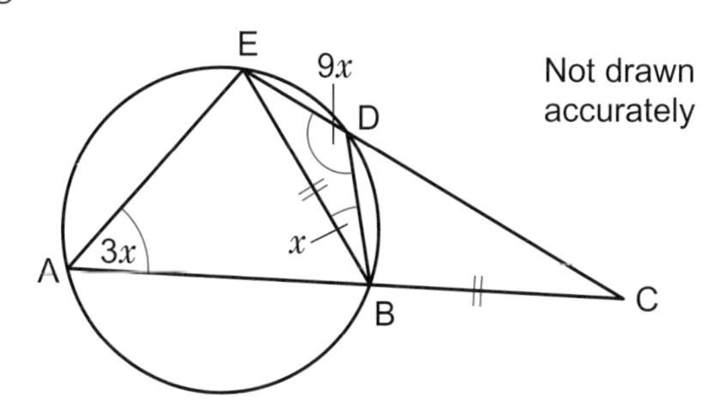

Work out the size of angle DBC, giving reasons for any statements you make.

Q9 A common error is to incorrectly name the angle being calculated in the workings.

10 Reasoning In the diagram, PQR is a straight line and PQ : QR = 2 : 3.

$\overrightarrow{OP} = 2\mathbf{a} + 5\mathbf{b}$

$\overrightarrow{OQ} = 4\mathbf{a} + \mathbf{b}$

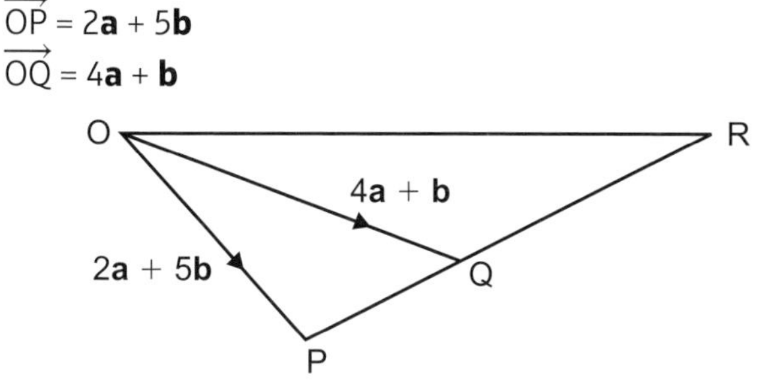

Write $\overrightarrow{OR}$ in terms of **a** and **b**. Give your answer in its simplest form.

Q10 A common error is to misunderstand ratio notation. If PQ : QR = 2 : 3, then PQ = $\frac{2}{5}$PR, not PQ = $\frac{2}{3}$PR. This is because PQ : QR = 2 : 3 means that PR is split into 5 parts.

11 Exam question

Jerry wants to cover a triangular field, ABC, with fertiliser.

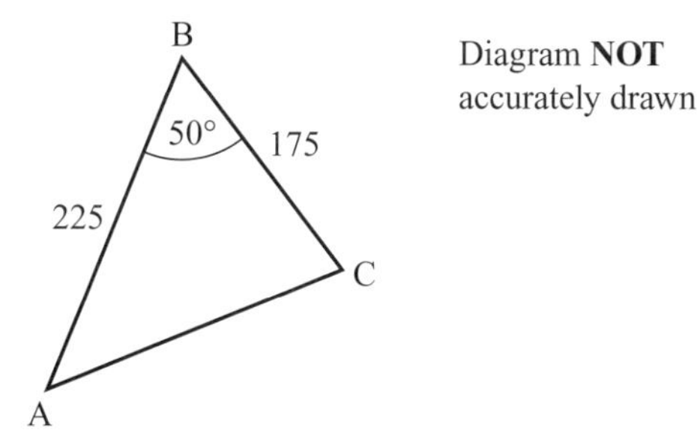

Diagram **NOT** accurately drawn

Here are the measurements Jerry makes:

angle ABC = 50° correct to the nearest degree

BA = 225 m correct to the nearest 5 m

BC = 175 m correct to the nearest 5 m.

Work out the upper bound for the area of the field.

You must show your working. **(3 marks)**

June 2014, Q20, 5MB3H/01

Exam hint

Some students correctly used the upper bounds for the lengths BA and BC but did not use the upper bound for sin 50°. Remember to use the upper bounds for all parts of the calculation.

12 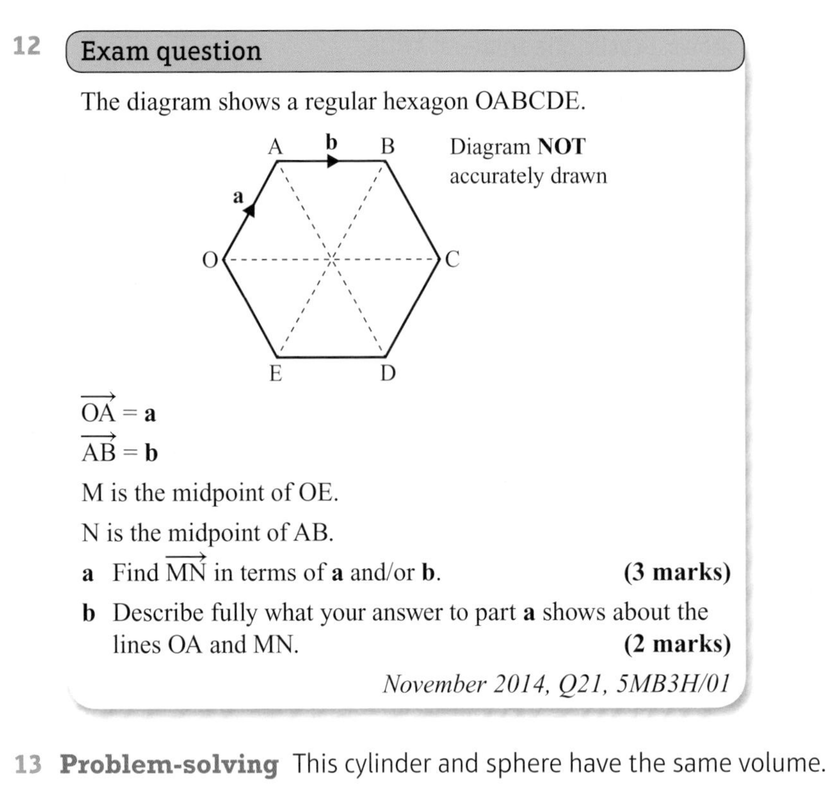

Exam question

The diagram shows a regular hexagon OABCDE.

$\overrightarrow{OA} = \mathbf{a}$

$\overrightarrow{AB} = \mathbf{b}$

M is the midpoint of OE.

N is the midpoint of AB.

a Find $\overrightarrow{MN}$ in terms of **a** and/or **b**. **(3 marks)**

b Describe fully what your answer to part **a** shows about the lines OA and MN. **(2 marks)**

November 2014, Q21, 5MB3H/01

13 **Problem-solving** This cylinder and sphere have the same volume.

x

$2y$

$3y$

Not drawn accurately

This cone and sphere have the same volume.

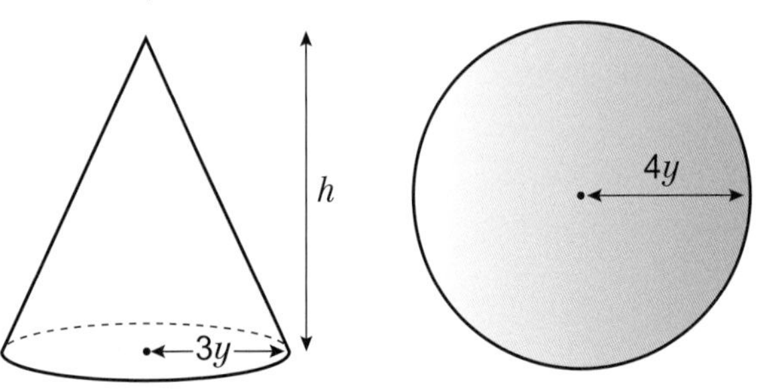

Not drawn accurately

Work out an expression for h in terms of x.

> **Q13** A common error is to not lay out working clearly, leading to careless errors.

14 **Problem-solving** OA = OB = 8.3 cm
Chord AB = 14 cm

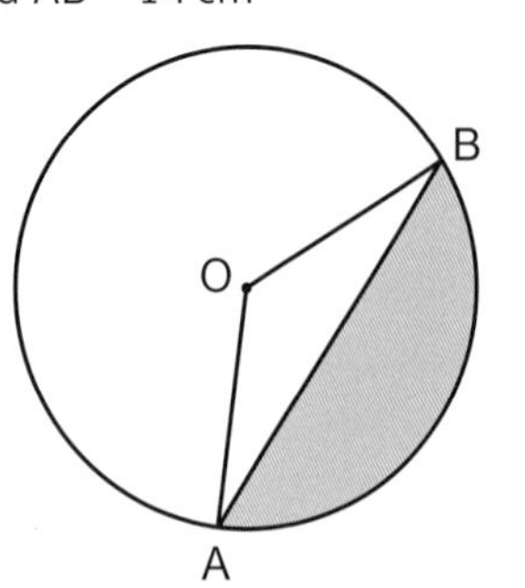

Work out the area of the shaded segment. Give your answer correct to 3 s.f.

15 Reasoning Work out the length of CD.
Give your answer correct to 3 s.f.

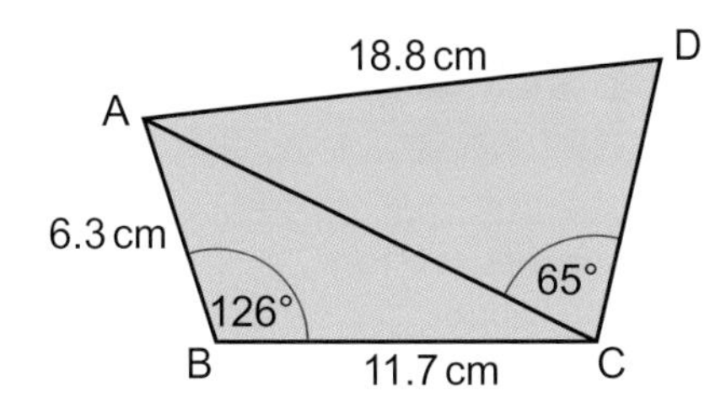

16 Reasoning The diagram shows a prism whose cross-section is a sector of a circle of radius 12 cm. The angle of the sector is 75° and the prism is 20 cm long.

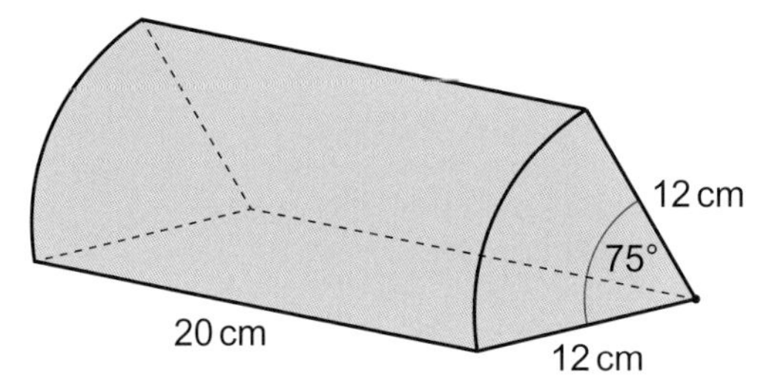

Show that the total surface area of the prism is $160(\pi + 3)$ cm².

17 Problem-solving OAPB is a parallelogram with $\overrightarrow{OA} = 6\mathbf{a}$ and $\overrightarrow{OB} = 4\mathbf{b}$.
BP is extended to Q such that $\overrightarrow{PQ} = 3\mathbf{a}$.
R lies on AB such that AR : RB = 2 : 3.

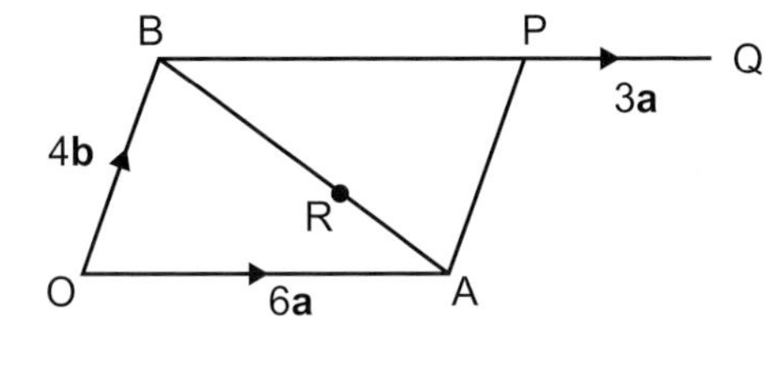

a Write down, in terms of **a** and **b**
  i $\overrightarrow{AB}$ ii $\overrightarrow{AR}$ iii $\overrightarrow{OR}$ iv $\overrightarrow{OQ}$

b What can you deduce about points O, R and Q? Show working to justify your answer.

18 Problem-solving The diagram shows a prism.
The cross-section is a right-angled triangle.
All of the other faces are rectangles.
AB = 12 cm
BC = 9 cm
CD = 16 cm
Calculate

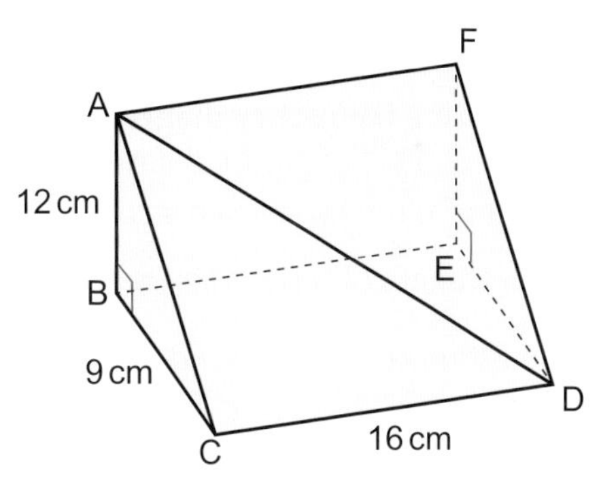

a the length of AD
b the angle that AD makes with the plane BCDE
c the angle that AD makes with the plane ABEF.

Give all your answers correct to 3 s.f.

19 The diagram shows a triangular-based pyramid.
All the faces are equilateral triangles, so the shape is a regular tetrahedron.
All the edges of the tetrahedron are 12 cm.
The perpendicular height, DN, is $4\sqrt{6}$ cm.
M is the midpoint of BC.
DM = $6\sqrt{3}$ cm
Calculate, giving your answers to 3 s.f.

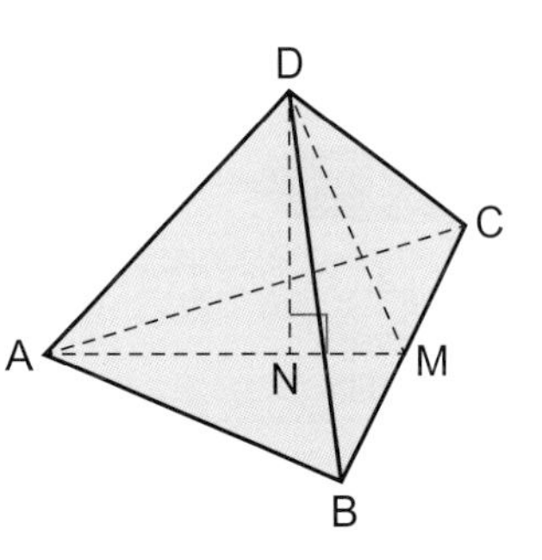

a the volume of the pyramid
b the total surface area of the pyramid
c the angle that DM makes with the base ABC.

6 STATISTICS

What is covered in this section	Higher Student Book reference
6.1 Comparison of distributions • Interpret, analyse and compare distributions using cumulative frequency diagrams and box plots, quoting a measure of average and a measure of spread.	Unit 14
6.2 Histograms • Use and understand frequency density, and construct and interpret histograms with unequal class intervals. • Complete a grouped frequency table from a histogram, determine frequency density from a histogram and estimate the median and other key values from a histogram with unequal class intervals.	Unit 14
6.3 Mixed exercise • Consolidate your learning with more practice.	

6.1 Comparison of distributions

Objective

- Interpret, analyse and compare distributions using cumulative frequency diagrams and box plots, quoting a measure of average and a measure of spread.

Key point 1

A **cumulative frequency diagram** can be used to estimate the median and the lower and upper quartiles of a set of grouped data.

For a set of n data values on a cumulative frequency diagram

- the estimate of the *median* is the $\frac{n}{2}$th value
- the estimate of the *lower quartile* (LQ) is the $\frac{n}{4}$th value
- the estimate of the *upper quartile* (UQ) is the $\frac{3n}{4}$th value
- the *interquartile range* (IQR) = UQ – LQ

These values are estimates because when data is grouped some of the detail of the original data is lost.

Key point 2

A **box plot** can be used to identify key features of a distribution.

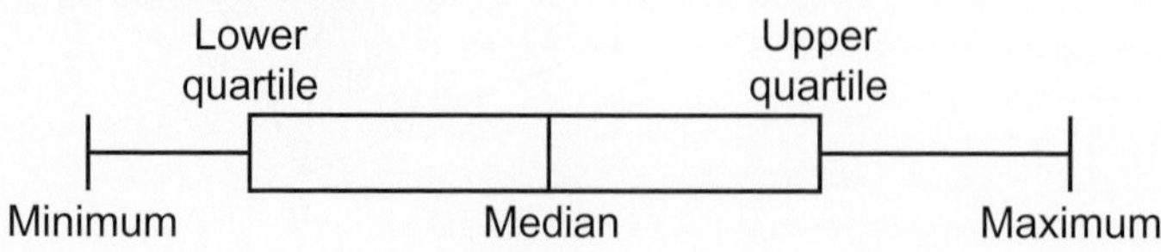

Box plots provide information 'at a glance' and give an impression of the general shape of a distribution but give no details about the data values themselves.

Key point 3

Box plots and cumulative frequency diagrams can be used to **compare** different sets of data. The median is a measure of average and the IQR measures the spread of the middle 50% of the data. These are reliable values to use because they are not affected by **outliers**.

1 a Work out the median of these numbers.
7, 30, 5, 19, 50, 15, 6, 32, 5, 15, 21, 9, 44
b The number 50 is removed. What happens to the median?

2 Write down six different numbers with a mean of 10.5 and a range of 8.

3 This frequency table gives information about the marks obtained by some students in a mental arithmetic test.

Mark	3	4	5	6	7	8	9	10
Frequency	2	1	4	1	3	5	3	1

Work out
a the mode b the median c the range d the mean.

4 120 people were timed on how long they took to complete a puzzle.
The cumulative frequency diagram shows the results.
a Estimate
i the median time ii the IQR.
b The shortest time taken was 12 minutes and the longest time was 37 minutes.
Draw a box plot to illustrate this data.

Q4 A common error is to make mistakes reading the scale on graphs.

Q4b hint Use the same scale as on the cumulative frequency diagram.

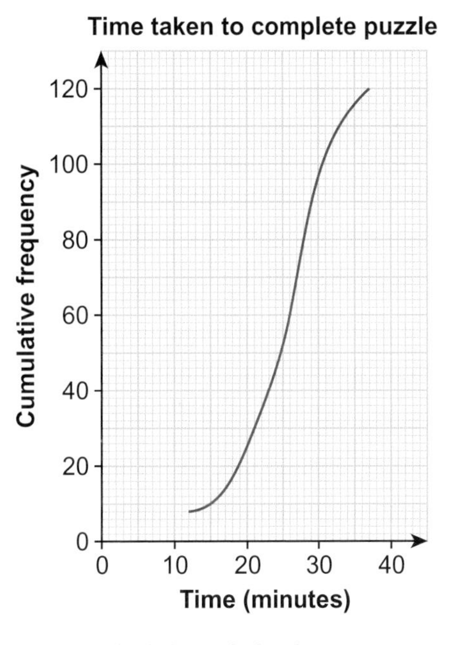

5 **Reasoning** The cumulative frequency diagram shows the heights of 80 shrubs at a garden centre.
a The smallest shrub was 18 cm and the tallest was 55 cm.
Draw a box plot to illustrate this data.
b What percentage of the shrubs were taller than 45 cm?
Did you use the cumulative frequency diagram or the box plot to obtain your answer? Give a reason for your choice.

Q5b A common error is to work out the percentage of shrubs smaller than 45 cm instead of taller than 45 cm. Read the question carefully.

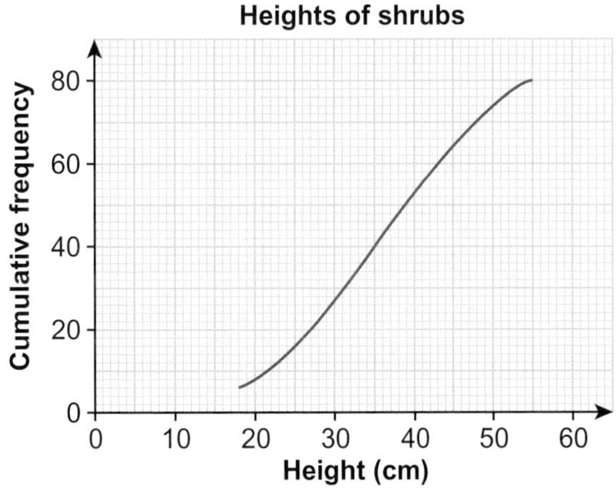

Warm up

6 **Reasoning** The box plots show the numbers of hours spent doing homework in one week by 32 boys and 26 girls.

a Give two differences between the amount of time spent by the boys and the amount of time spent by the girls.

b How many students did more than 7.6 hours of homework?

Q6b hint The UQ for the boys is 7.6 hours. What proportion of the boys spent longer than 7.6 hours?

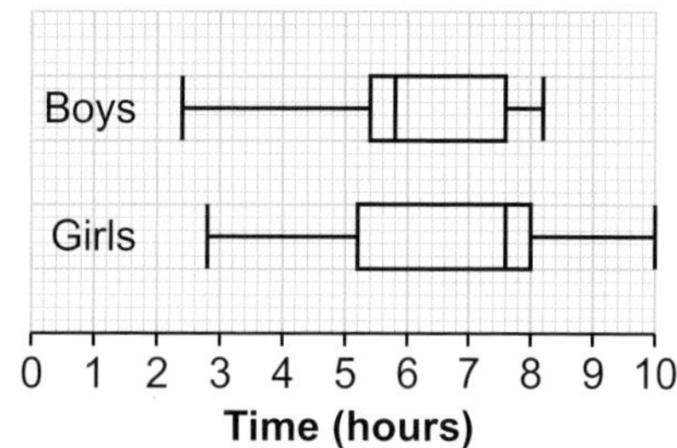

7 **Problem-solving** Tomatoes of two different varieties, X and Y, are weighed. The cumulative frequency diagram shows the frequency distributions of the masses of the two varieties.

a How many tomatoes of variety X were weighed?

b Estimate the minimum and maximum masses of tomatoes from variety X.

c Work out the median and the interquartile range for tomatoes from variety X.

d How many tomatoes of variety Y were weighed?

e Estimate the minimum and maximum masses of tomatoes from variety Y.

f Work out the median and the interquartile range for tomatoes from variety Y.

g Draw box plots for both varieties using a scale from 150 to 210 on the horizontal axis.

h Compare the two varieties, giving reasons for any statements you make.

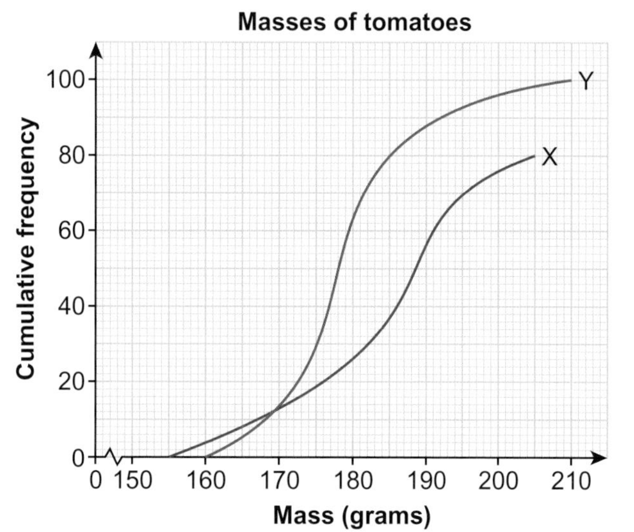

Q7h hint Your answer should be phrased within the context of the question and should refer to masses of tomatoes.

8 **Problem-solving** The table shows the numbers of days 100 people had to wait to get a driving test in town A.

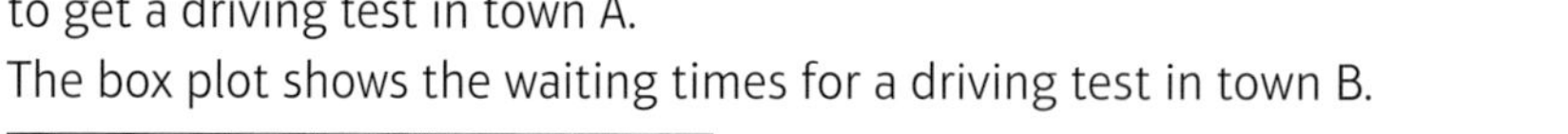

The box plot shows the waiting times for a driving test in town B.

Number of days, d	Frequency
$10 < d \leqslant 20$	2
$20 < d \leqslant 30$	9
$30 < d \leqslant 40$	22
$40 < d \leqslant 50$	34
$50 < d \leqslant 60$	23
$60 < d \leqslant 70$	10

Compare the waiting times for a driving test in towns A and B.

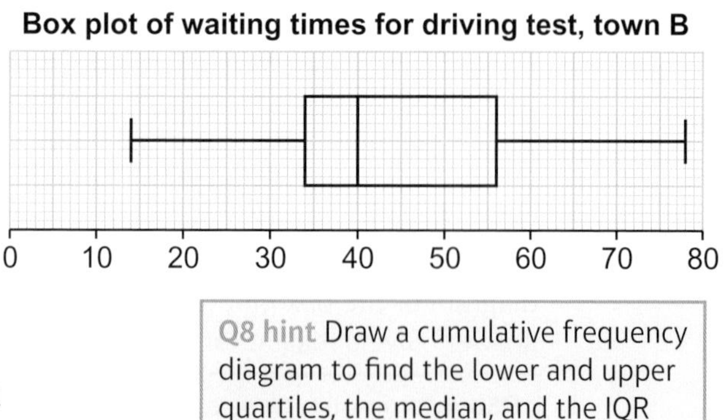

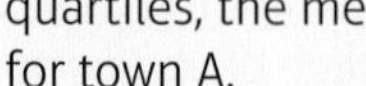

Q8 hint Draw a cumulative frequency diagram to find the lower and upper quartiles, the median, and the IQR for town A.

6.2 Histograms

Objectives

- Use and understand frequency density, and construct and interpret histograms with unequal class intervals.
- Complete a grouped frequency table from a histogram, determine frequency density from a histogram and estimate the median and other key values from a histogram with unequal class intervals.

Key point 4

A **histogram** is a diagram that represents continuous data. It is similar to a bar chart except that

- the data is continuous so there are no gaps between the bars
- the bars can be different widths to represent different class intervals
- the areas of the bars are proportional to the frequencies they represent.

Key point 5

For a histogram, the scale on the horizontal axis should be continuous. The vertical axis shows the **frequency density** and always starts at zero.

The frequency density is given by the formula

$$\text{frequency density} = \frac{\text{frequency}}{\text{class width}}$$

and hence

frequency = frequency density × class width

1 The table shows the heights of 40 people attending the gym at their local leisure centre.

Height, h (cm)	$160 < h \leqslant 165$	$165 < h \leqslant 170$	$170 < h \leqslant 175$	$175 < h \leqslant 180$	$180 < h \leqslant 185$
Frequency	12	10	9	6	3

a Draw a frequency diagram for this data.
b Write down the modal class.
c Which class interval contains the median?
d Work out an estimate of the mean height.

2 The table shows the ten-pin bowling scores for 300 people in the local league.

Score, s	**Frequency**
$75 < s \leqslant 125$	45
$125 < s \leqslant 150$	50
$150 < s \leqslant 175$	70
$175 < s \leqslant 225$	90
$225 < s \leqslant 300$	45

Draw a fully labelled histogram to represent the data.

Q2 hint Frequency density = $\frac{\text{frequency}}{\text{class width}}$
The scale on the frequency density axis starts at zero.

Q2 A common error is to use class intervals, not a linear scale, on the horizontal axis.

Warm up

3 **Reasoning** The histogram shows the test scores of 320 children in a school.

a Estimate the median test score.

b Work out an estimate of the interquartile range of the scores.

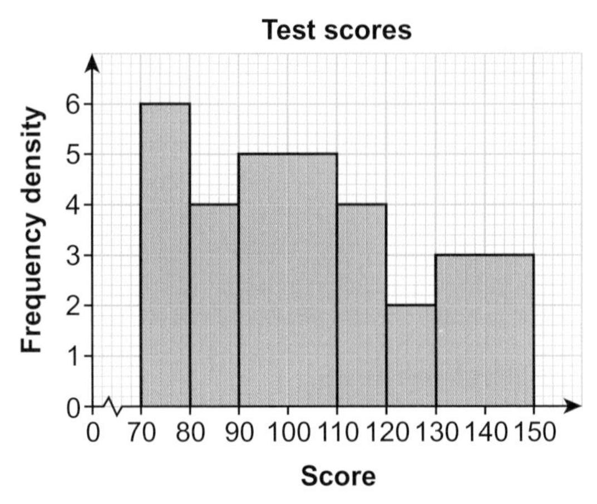

Q3 hint Work out which classes contain the median and the quartiles, then locate their positions within these classes.

4 **Reasoning** The histogram shows patients' waiting times in a doctors' surgery during one week.

a Rachel says, 'More than half the patients waited less than 20 minutes.'
Is Rachel correct?
Explain your answer.

b Estimate the number of patients who waited for more than 33 minutes.

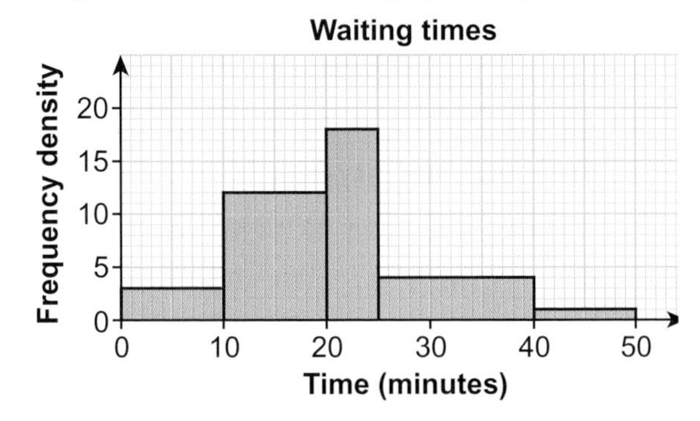

Q4a A common error is to use the heights of the bars instead of the areas.

5 **Problem-solving** The histogram and the frequency table show the same information about the times that vehicles spent in a car park.

Time, t (minutes)	Frequency
$0 < t \leqslant 30$	45
$30 < t \leqslant 60$	54
$60 < t \leqslant 100$	
$100 < t \leqslant 120$	50
$120 < t \leqslant 180$	30

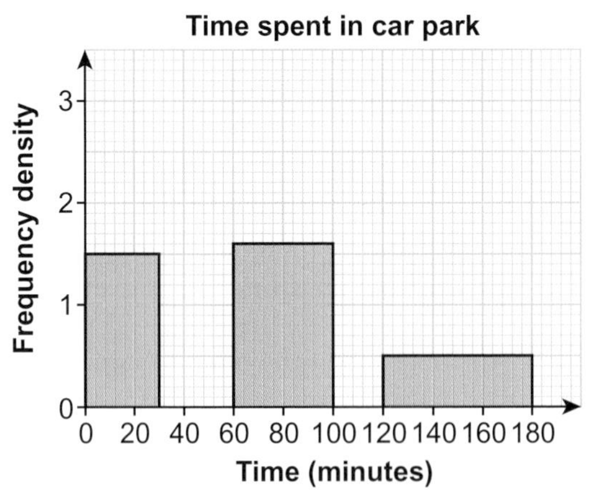

a Copy and complete the histogram and work out the missing number in the frequency table.

b 50 vehicles were in the car park for more than T minutes.
Work out an estimate of the value of T.

6 The histogram represents the heights of plants, in centimetres, at a garden centre.

a How many plants are represented by the histogram?

b Estimate the median height of the plants.

c Estimate how many plants are more than 36 cm tall.

Q6b A common error is to give the median as the middle of the range 5–45 or as the middle value of the $20 < n \leqslant 30$ class.

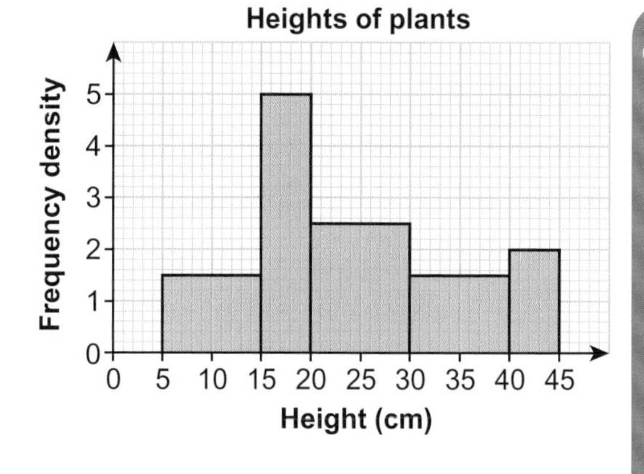

7 **Problem-solving** The histogram shows the distribution of the masses of 80 young finches.

a Estimate the median mass of a young finch.

b Compile a grouped frequency table using the information in the histogram.

c Use your grouped frequency table to calculate an estimate of the mean mass of a young finch.

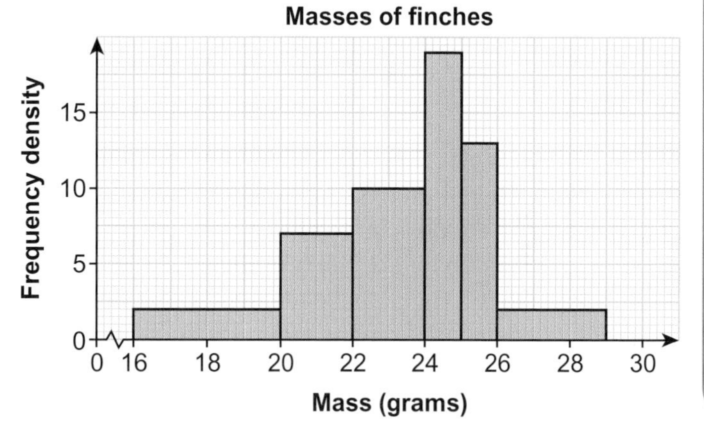

8 **Problem-solving** The histogram shows the finishing times of the runners in a 10 km cross-country race.

a How many runners were in the race?

b There was a target time of 67 minutes. How many runners failed to meet the target time?

c i Compile a grouped frequency table using the information in the histogram.

ii Use your table to calculate an estimate of the mean time.

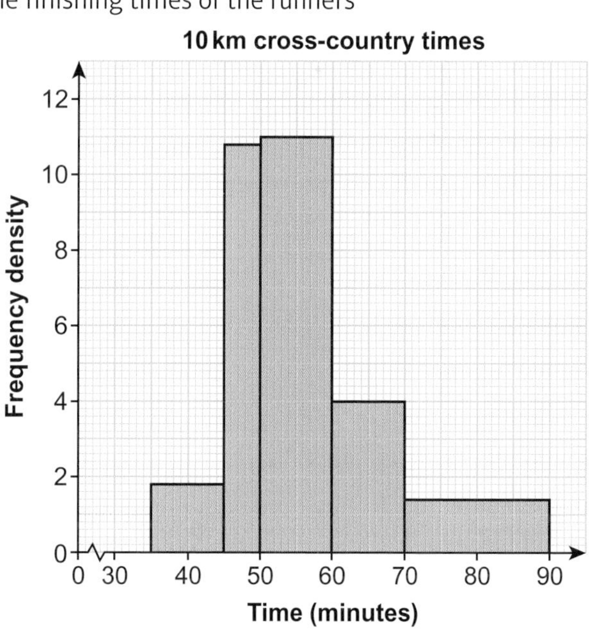

6.3 Mixed exercise

Objective

- Consolidate your learning with more practice.

1 Paula is planning a holiday. She looks at temperature data for two destinations, A and B. The box plots show information about the summer temperatures in the two places.

a Give *two* reasons why Paula might choose destination A.

b Give *two* reasons why Paula might choose destination B.

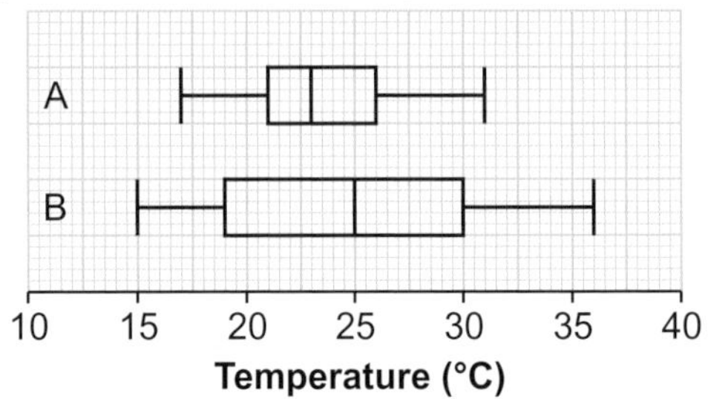

2 The box plot shows the marks of 40 students in a maths test. Two students scored the minimum mark of 4.

a Use the box plot to draw a cumulative frequency diagram for the marks of the 40 students.

b What is the probability that a student picked at random from the group scored more than 34?

Q2a hint Use the same horizontal scale as on the box plot.

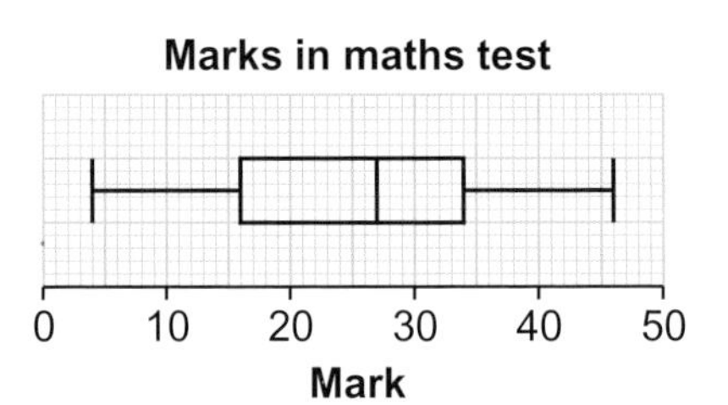

3 **Reasoning** The cumulative frequency diagram shows the weights at birth of 80 babies born in a hospital in Europe.

a How many babies have a birthweight of 3.5 kg or less?

b What is the median birthweight?

c Work out the interquartile range of the birthweights.

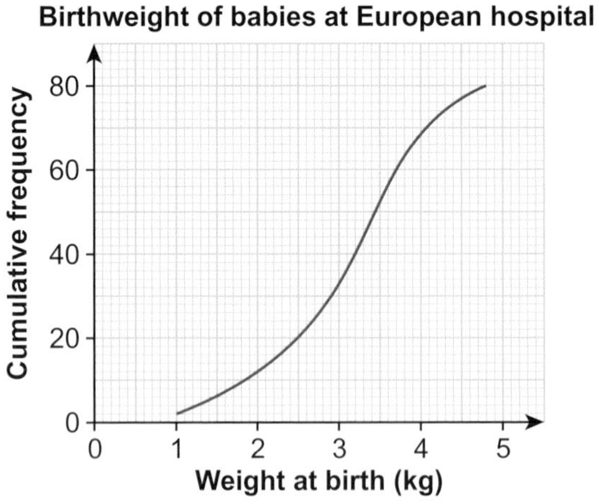

d The box plot shows the weights at birth of 80 babies born in a hospital in Asia. Use the data to give *two* comparisons between the birthweights of babies in Europe and Asia.

Q3d hint You could draw a box plot for the Europe data, and then compare the two box plots.

Birthweight of babies at Asian hospital

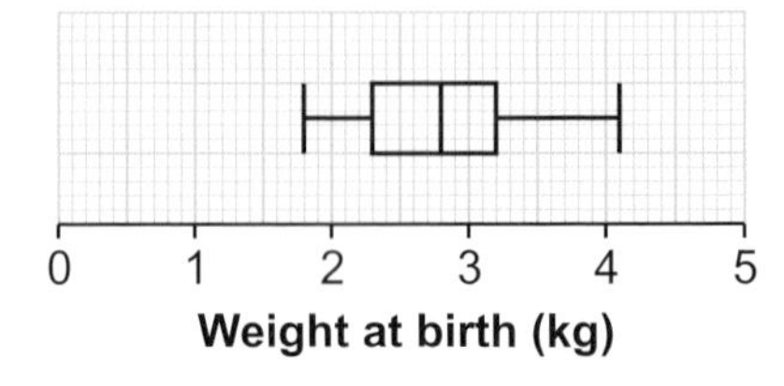

4

Exam question

The cumulative frequency graph shows information about the times 80 swimmers take to swim 50 metres.

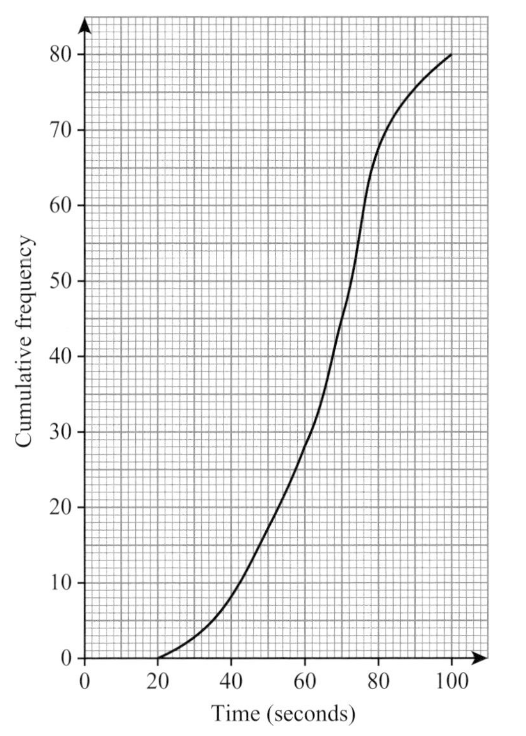

a Use the graph to find an estimate for the median time. **(1 mark)**

A swimmer has to swim 50 metres in 60 seconds or less to qualify for the swimming team.

The team captain says,

"More than 25% of swimmers have qualified for the swimming team."

b Is the team captain right? You must show how you got your answer. **(3 marks)**

For these 80 swimmers

the least time taken was 28 seconds

and the greatest time taken was 96 seconds.

c Use the cumulative frequency graph and the information above to draw a box plot for the times taken by the swimmers.

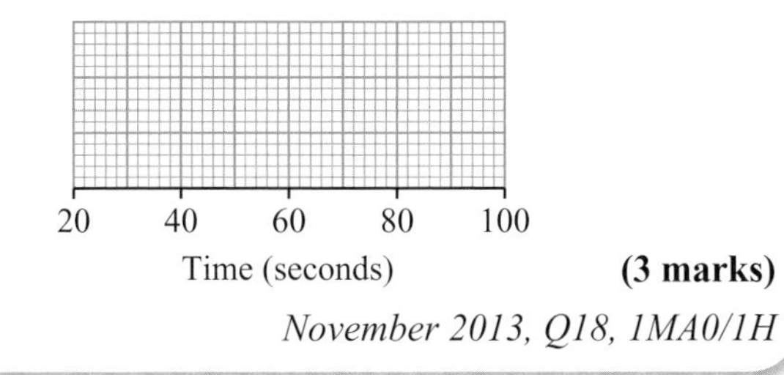

(3 marks)

November 2013, Q18, 1MA0/1H

Q4b A common error is to simply write 'Yes' or 'No' as the answer. You must give reasons to justify your answer.

5 **Exam question**

The table gives information about the heights, h, of trees in a wood.

Height, h (m)	Frequency
$0 < h \leq 2$	7
$2 < h \leq 4$	14
$4 < h \leq 8$	18
$8 < h \leq 16$	24
$16 < h \leq 20$	10

Draw a histogram to show this information. **(3 marks)**

November 2012, Q24, 1MA0/2H

Q5 A common error is to use frequency instead of frequency density on the y-axis.

6 **Problem-solving** A speed camera recorded the speeds of 400 vehicles on a motorway. The frequency table shows the results.

a Draw a histogram to illustrate this data.

b Drivers of vehicles travelling faster than 77 mph were given a speeding ticket. Estimate the number of drivers who received a ticket.

Speed, x (mph)	Frequency
$0 < x \leq 30$	36
$30 < x \leq 50$	68
$50 < x \leq 60$	86
$60 < x \leq 70$	98
$70 < x \leq 80$	80
$80 < x \leq 120$	32

7 **Problem-solving** The table shows the times that vehicles spent in an out-of-town shopping centre car park.

Time, t (minutes)	Frequency
$0 < t \leq 30$	18
$30 < t \leq 60$	
$60 < t \leq 150$	180
$150 < t \leq 210$	90
$210 < t \leq 240$	57

The histogram shows the same information.

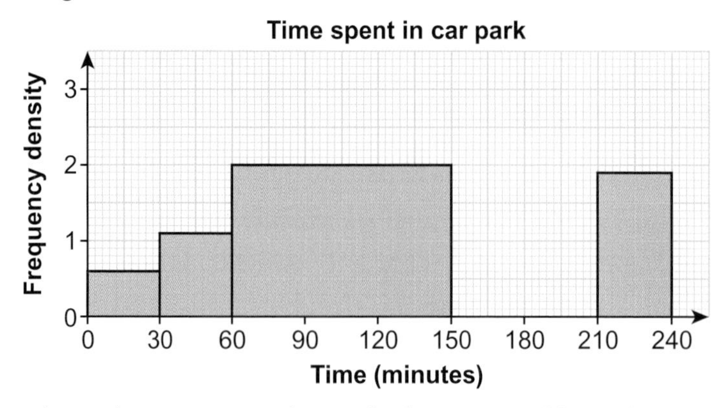

a Work out the missing number in the frequency table, and copy and complete the histogram.

b Estimate how many vehicles were in the car park for 120–180 minutes.

8 **Exam question**

The histogram gives information about the areas of 285 farms.

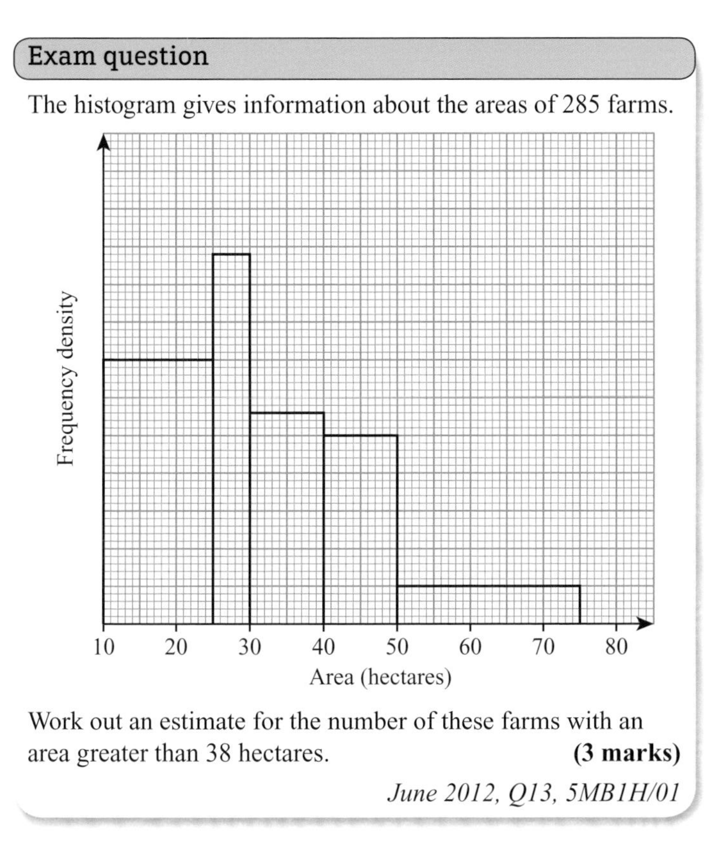

Work out an estimate for the number of these farms with an area greater than 38 hectares. **(3 marks)**

June 2012, Q13, 5MB1H/01

9 **Problem-solving / Reasoning** The frequency table and the histogram show the same data for the heights of 104 adults.

Height, h (cm)	Frequency
$140 < h \leqslant 160$	22
$160 < h \leqslant 165$	
$165 < h \leqslant 170$	
$170 < h \leqslant 185$	27
$185 < h \leqslant 200$	12

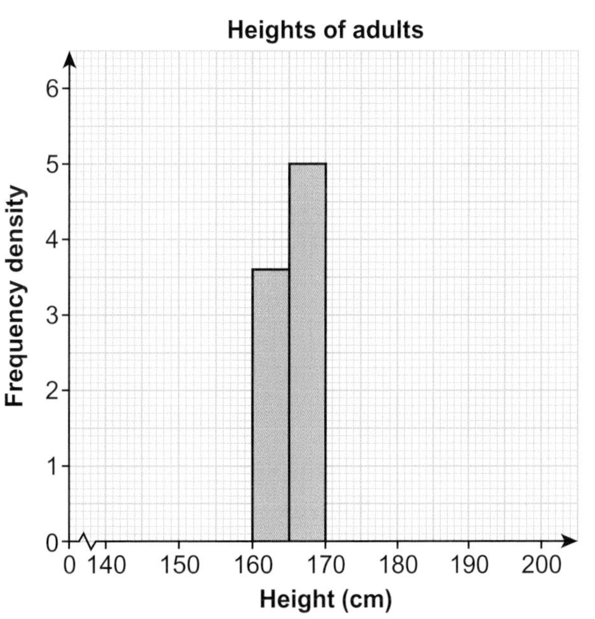

a Copy and complete the frequency table and the histogram.

b Clare says that an estimate of the median height will be between 167 cm and 168 cm. Is she correct? Show working to justify your answer.

c Calculate an estimate of the mean height. Give your answer correct to 1 d.p.

ANSWERS

1 NUMBER

1.1 Accuracy and bounds

1 a 23.5 cm $\leqslant x <$ 24.5 cm
b 122.5 m $\leqslant x <$ 127.5 m
c 9.75 seconds $\leqslant x <$ 9.85 seconds
d 8.455 kg $\leqslant x <$ 8.465 kg

2 a lb = 5.5 m, ub = 6.5 m
b lb = 65 mm, ub = 75 mm
c lb = 4.85 litres, ub = 4.95 litres
d lb = 19.755 seconds, ub = 19.765 seconds

3 199.5 cm − 83.55 cm = 115.95 cm

4 2(106.5 + 78.5) m = 370 m

5 lb = 8.45 × 5.15 = 43.5175 cm^2,
ub = 8.55 × 5.25 = 44.8875 cm^2

6 Distance = speed × time = 23.65 × 18.5 = 437.525 m

7 Volume = 7.153 = 365.525 875 cm^3

8 Maximum weight = 16.5 + 12 × 67.5 = 826.5 g.
Minimum weight = 15.5 + 12 × 66.5 = 813.5 g.
Difference = 13 g

9 a 2 × 23.55 − 3 × 9.45 = 18.75
b 18.75^2 − 23.55 × 9.35 = 131.37

10 Density = mass ÷ volume
Minimum density = 238.5 ÷ 12.45 = 19.1566…
Maximum density = 239.5 ÷ 12.35 = 19.3927…
19.16 $\leqslant d <$ 19.39

11 Area = $\frac{1}{2}$ × base × height
Minimum height = 2 × minimum area ÷ maximum base
= 2 × 47.5 ÷ 14.35 = 6.62 cm

12 Minimum value = $\dfrac{225}{(8.15^2 - 13.85)} = 4.28$
Maximum value = $\dfrac{235}{(8.05^2 - 13.95)} = 4.62$

1.2 Number problems and reasoning

1 a 10^4 = 10 000 b 10 × 9 × 8 × 7 = 5040

2 a 12 × 60 = 720
b 12 × 12 = 144
c 16 × 125 = 2000

3 (26 × 25 × 24) × (10 × 9 × 8) × 26 = 292 032 000

4 a 6 × 5 × 4 = 120
b 4 × 5 × 4 = 80
c 1 × 5 × 4 = 20

5 a 4 × 6 × 5 = 120 b (4 × 6) + (6 × 5) = 54

6 (6 × 5) ÷ 2 = 15

7 3 × 2 × 6! = 4320

8 a 3 × 6 × 5 = 90
b (1 × 2 × 5) + (1 × 3 × 5) = 10 + 15 = 25

1.3 Indices, roots and surds

1 a 128 b 1 c 243 d 125
e $\frac{1}{256}$ f 625 g 1000 h 256

2 a $x^2 + 5x - 24$ b $3x^2 - 11x + 10$
c $4x^2 + 28x + 49$ d $4x^2 - 7xy - 2y^2$

3 a 2 b −6 c 512 d 125

4 a $\frac{1}{9}$ b $\frac{1}{125}$ c $\frac{1}{7}$ d $-\frac{4}{9}$
e 8

5 a $\frac{49}{16}$ b $\frac{27}{64}$ c $\frac{32}{243}$ d $\frac{81}{625}$

6 a $5\sqrt{3}$ b $12\sqrt{2}$ c $15\sqrt{2}$ d $3\sqrt{5}$
e $8\sqrt{7}$ f $5\sqrt{11}$

7 a 6 b 10 c 3 d 15

8 a 12 b $6\sqrt{5}$ c $20\sqrt{2}$ d 10.5

9 a $4\sqrt{3}$ b 90 c 72 d $70\sqrt{5}$

10 a $6\sqrt{3}$ b 20 c $2\sqrt{3}$ d 15
e $140\sqrt{5}$ f $40\sqrt{2}$

11 a $4 + \sqrt{3}$ b $23 + 8\sqrt{7}$ c $6\sqrt{6} - 14$

12 a $7 + 6\sqrt{2}$ b $18 + 11\sqrt{3}$ c $33 - 20\sqrt{2}$
d $109 - 48\sqrt{5}$ e $13 + 25\sqrt{2}$ f 16

13 a 7 b 94 c −8 d 18

14 10

15 $2^2 + (2 + \sqrt{3})^2 = 4 + 7 + 4\sqrt{3} = 11 + 4\sqrt{3}$ and
$(1 + \sqrt{12})^2 = 13 + 4\sqrt{3}$. Triangle is not right-angled.

16 a $7\sqrt{2}$ b $6\sqrt{3}$ c $\dfrac{\sqrt{10}}{2}$ d $4\sqrt{3}$
e $\dfrac{3\sqrt{5}}{2}$ f $\dfrac{\sqrt{3}}{6}$

17 $\dfrac{\sqrt{3}}{10}$

18 a $3(\sqrt{2} - 1) = 3\sqrt{2} - 3$ b $4(\sqrt{5} + 2) = 4\sqrt{5} + 8$
c $5(\sqrt{5} - \sqrt{3}) = 5\sqrt{5} - 5\sqrt{3}$ d $2(3 - \sqrt{2}) = 6 - 2\sqrt{2}$
e $\dfrac{5 - \sqrt{7}}{2} = \dfrac{5}{2} - \dfrac{\sqrt{7}}{2}$ f $\dfrac{2(6 + \sqrt{3})}{3} = 4 + \dfrac{2\sqrt{3}}{3}$

1.4 Mixed exercise

1 a ub = 21.45 seconds, lb = 21.35 seconds
b ub = 37.5 cm, lb = 32.5 cm
c ub = 1435 kg, lb = 1425 kg
d ub = 16.285 litres, lb = 16.275 litres

2 72.5 ÷ 6.75 = 10.74 cm

3 a 4 × 3 × 3 = 36 b 2 × 3 × 2 = 12

4 a 7 b $\frac{1}{10}$ c 2

5 a $6\sqrt{2}$ b 240 c $\dfrac{5\sqrt{2}}{8}$

6 Minimum value = $\dfrac{9.15^2}{3 \times 8.65 - 5.35} = 4.06$
Maximum value = $\dfrac{9.25^2}{3 \times 8.55 - 5.45} = 4.24$

7 18 cm^2

8 a 32 b 216 c $\frac{1}{625}$

9 $9 - \sqrt{5}$

10 a $\dfrac{8\sqrt{3}}{5}$ b $\dfrac{8 + \sqrt{7}}{3}$

11 a $\frac{36}{25}$ b $\frac{343}{512}$ c $\frac{243}{100\,000}$ d $\frac{27}{8}$

2 ALGEBRA

2.1 Parallel and perpendicular lines

1 a 1 b $-\frac{1}{2}$ c 3.5

2 a −2 b $y = -2x + 10$

3 a Lines B and D b Lines C and E

4 a 3 b $-\frac{1}{3}$ c $y = -\frac{1}{3}x + 1$

5 $y = \frac{2}{3}x - 5$

6 (15, 0)

2.2 Quadratic functions

1 a i $x = -1.5$
ii (−1.5, −2.25), minimum
iii $x = -3$ or 0
b i $x = 0.5$
ii (0.5, −4.25), minimum
iii $x = -1.56$ or 2.56
c i $x = -1$
ii (−1, 6), maximum
iii $x = -3.45$ or 1.45
d i $x = -0.75$
ii (−0.75, −2.13), minimum
iii $x = -1.78$ or 0.28

2 a

x	−3	−2	−1	0	1	2	3	4
y	9	3	−1	−3	−3	−1	3	9

b Graph of $y = x^2 - x - 3$ plotted using information in the completed table.
c $x = -1.3$ or 2.3
d Line $y = x + 1$ added to graph; $x = -1.2$ or 3.2

3 a

x	−1	0	1	2	3	4
y	8	1	−2	−1	4	13

b Graph of $y = 2x^2 - 5x + 1$ plotted using information in the completed table.
c $x = 0.2$ or 2.3
d Line $y = 4 - 2x$ added to graph; $x = -0.7$ or 2.2

4 a $x^2 - 5x - 24$ b $2x^2 - 13x + 20$
c $10x^2 + 31x + 15$ d $6x^2 + 11x - 7$
e $3x^2 + 2xy - 8y^2$ f $12x^2 - 17xy + 6y^2$

5 a $(x + 3)(x - 3)$ b $(6 + 5y)(6 - 5y)$
c $5(x + 2)(x - 2)$ d $3(2 + 3w)(2 - 3w)$
e $5(3c + 2d)(3c - 2d)$ f $2(4m + 5t)(4m - 5t)$

6 a $x = \frac{3}{2}$ or −5 b $x = \frac{3}{2}$ or 5
c $x = -\frac{7}{2}$ or 4 d $x = -\frac{3}{5}$ or −7
e $x = \frac{1}{4}$ or −4 f $x = -\frac{9}{2}$ or $\frac{3}{2}$

7 a $x = 3.41$ or 0.59 b $y = 1.74$ or −5.74
c $w = 1.85$ or −1.35 d $x = 2.47$ or −0.14
e $y = 1.22$ or −0.82 f $w = -0.19$ or −1.31

8 a 2 roots (discriminant = 32)
b 0 roots (discriminant = −7)
c 2 roots (discriminant = 73)
d 1 root (discriminant = 0)
e 2 roots (discriminant = 157)
f 0 roots (discriminant = −4)

9 $m = 16$ or −16

10 $k = 3.2$

11 a $(x + 1)^2 + 9$ b $(x - 5)^2 + 5$
c $(x + 4)^2 - 17$ d $2(x + 1)^2 - 5$
e $3(x - 1)^2 - 2$ f $4(x - 2)^2 - 11$

12 a $-1 \pm \sqrt{8}$ b $-6 \pm \sqrt{40}$
c $4 \pm \sqrt{3}$ d $-3 \pm \sqrt{7}$
e $5 \pm \sqrt{5}$ f $2.5 \pm \sqrt{6}$

13 a (3, −5)

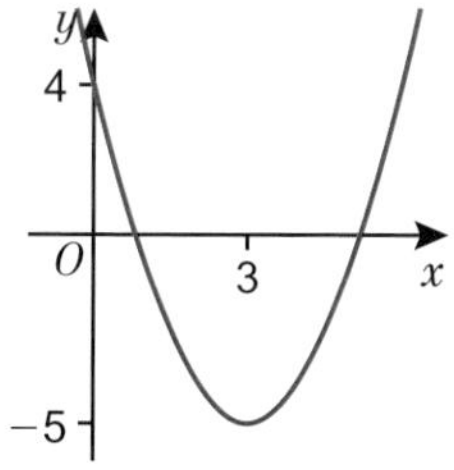

b (−1, −7)

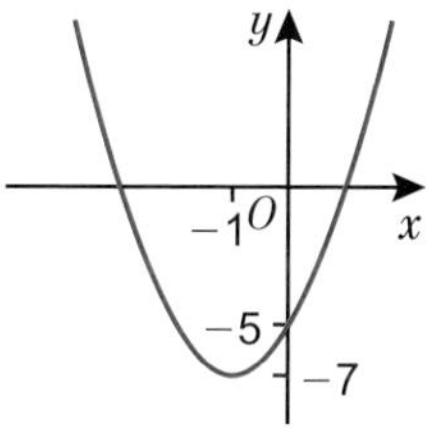

c (2, −5)

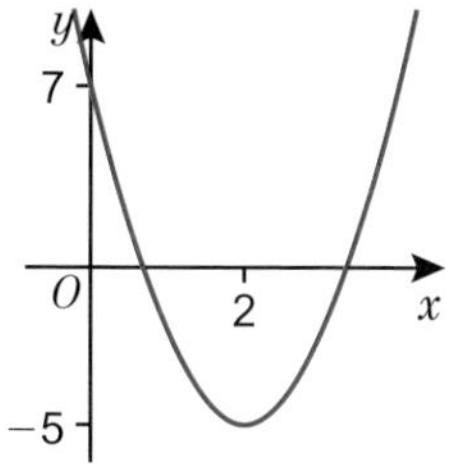

14 $2(x^2 - 3x) + 7 = 0$
$2[(x - 1.5)^2 - 2.25] + 7 = 0$
$2(x - 1.5)^2 - 4.5 + 7 = 0$
$2(x - 1.5)^2 = -2.5$
Since the RHS is negative there are no real roots.

2.3 Inverse and composite functions

1 a 1 b −11 c −47 d $\frac{1}{4}$

2 a $n = 2$ b $n = \frac{1}{2}$ c $n = \frac{7}{4}$ d $n = 4.5$

3 a $t = 3$ or −1 b $t = 5$ or −3 c $t = 10$ or −8

4 a $12x + 16$ b $48x + 20$ c $18x - 2$

5 a $m = -\frac{1}{2}$ or −1 b $m = -\frac{5}{2}$ or 1
c $m = 5$ or −2 d $m = -4$ or 7

6 a $5x^2 + 3$ b $20x^2 + 3$
c $5x^2 + 20x + 23$ d $45x^2 - 120x + 83$

7 a −15 b 116 c 5
d 76 e $21 - 4x^2$ f $16x^2 - 8x - 4$

8 a $33 - 5x$ b $9 - 5x$
c $1 - x^2$ d $x^2 - 14x + 55$
e $5x^2 + 28$ f $25x^2 - 20x + 10$

9 a $4x^2 - 10x - 6$ b $x = -\frac{1}{2}$ or 3

10 a $f^{-1}(x) = \frac{x + 4}{3}$ b $f^{-1}(x) = \frac{x}{3} + 4$
c $f^{-1}(x) = 2(x - 5)$ d $f^{-1}(x) = 2x - 5$
e $f^{-1}(x) = \frac{x + 15}{2}$ f $f^{-1}(x) = \frac{x - 10}{-3}$ or $\frac{10 - x}{3}$

2.4 Gradient and area under linear and non-linear graphs

1 a 120 miles from London b 80 mph, 0 mph, 114.3 mph
c $\frac{280}{3.25} = 86.2$ mph

2 a 7 m/s b 4 seconds c 3.5 m/s² d 47 m

3 a 5 m/s²
b 14 m/s constant velocity for 3 seconds.
Accelerates at 1 m/s² for 4 seconds.
18 m/s constant velocity for 6 seconds.
Deceleration of 5 m/s² for 2 seconds.
8 m/s constant velocity for 2 seconds.
Deceleration of $\frac{8}{3}$ m/s² for the final 3 seconds, bringing the car to rest.
c 268 m
d $\frac{268}{20} = 13.4$ m/s (≈ 30 mph)

4 At $x = -1$, gradient = −3
At $x = 0.5$, gradient = 0
At $x = 1$, gradient = 1

5 a

x	−6	−5	−4	−3	−2	−1	0	1
y	7	1	−3	−5	−5	−3	1	7

b Graph of $y = x^2 + 5x + 1$ plotted using information in the completed table.
c Gradient at $x = -2$ is 1.
d Gradient at $x = -4$ is −3.

6 a

x	−1	0	1	2	3	4	5
y	−2	3	6	7	6	3	−2

b Graph of $y = 3 + 4x - x^2$ plotted using information in the completed table.
c Gradient at $x = 1$ is 2.
d Gradient at $x = 4$ is −4.

7 a

x	−4	−3	−2	−1	0	1
y	2	−5	−8	−7	−2	7

b Graph of $y = 2x^2 + 7x - 2$ plotted using information in the completed table.
c Gradient at $x = -2.5$ is −3.
d Gradient at $x = -0.5$ is 5.

8 a $6\frac{1}{2}$ square units

b A = $2\frac{1}{2}$, B = $6\frac{1}{2}$, C = $8\frac{1}{2}$, D = $8\frac{1}{2}$, E = $6\frac{1}{2}$, F = $2\frac{1}{2}$

c 35 square units

d The area of each of the triangles and trapezia is slightly less than the area under the graph, owing to the areas lost by drawing chords. But these areas are very small, so 35 is a good approximation.

9 a

x	−2	−1	0	1	2	3
y	2	0	0	2	6	12

b Graph of $y = x^2 + x$ plotted using information in the completed table.

c 14 square units

d An overestimate, because the area of each trapezium is slightly more than the real area.

10 a

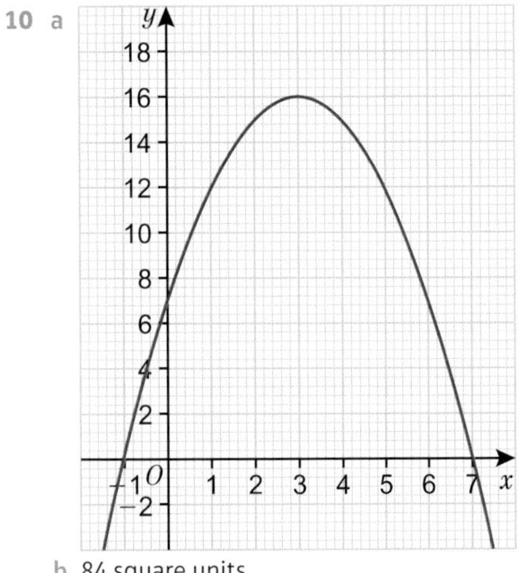

b 84 square units

11 a About $\frac{19}{50}$ or 0.38 m/s²

b About 0.24 m/s²

c About 7750 m

12 a About 6 m/s²

b About 297 m

c $T \approx 4.6$

2.5 The circle, $x^2 + y^2 = r^2$

1 a Not a circle

b Circle, radius 2

c Not a circle

d Not a circle

e Circle, radius 9

f Circle, radius $\sqrt{6}$

2 a $-\frac{1}{3}$ b $\frac{1}{4}$ c −2 d 5

e $-\frac{5}{4}$ f $-\frac{2}{7}$ g $\frac{2}{5}$ h $-\frac{19}{4}$

3 a

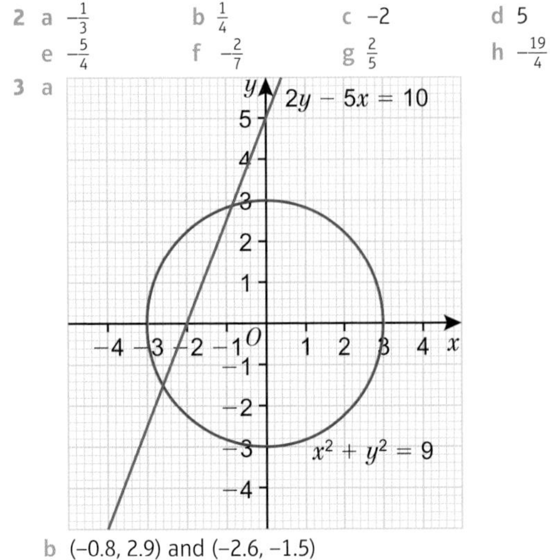

b (−0.8, 2.9) and (−2.6, −1.5)

4 a

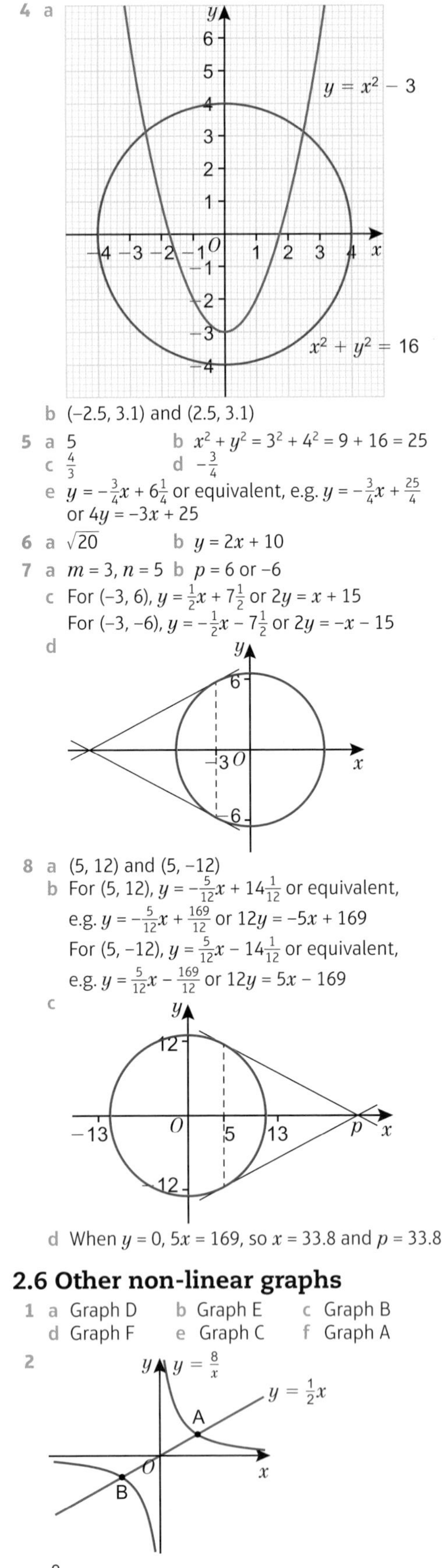

b (−2.5, 3.1) and (2.5, 3.1)

5 a 5 b $x^2 + y^2 = 3^2 + 4^2 = 9 + 16 = 25$

c $\frac{4}{3}$ d $-\frac{3}{4}$

e $y = -\frac{3}{4}x + 6\frac{1}{4}$ or equivalent, e.g. $y = -\frac{3}{4}x + \frac{25}{4}$ or $4y = -3x + 25$

6 a $\sqrt{20}$ b $y = 2x + 10$

7 a $m = 3, n = 5$ b $p = 6$ or -6

c For (−3, 6), $y = \frac{1}{2}x + 7\frac{1}{2}$ or $2y = x + 15$

For (−3, −6), $y = -\frac{1}{2}x - 7\frac{1}{2}$ or $2y = -x - 15$

d

8 a (5, 12) and (5, −12)

b For (5, 12), $y = -\frac{5}{12}x + 14\frac{1}{12}$ or equivalent, e.g. $y = -\frac{5}{12}x + \frac{169}{12}$ or $12y = -5x + 169$

For (5, −12), $y = \frac{5}{12}x - 14\frac{1}{12}$ or equivalent, e.g. $y = \frac{5}{12}x - \frac{169}{12}$ or $12y = 5x - 169$

c

d When $y = 0$, $5x = 169$, so $x = 33.8$ and $p = 33.8$

2.6 Other non-linear graphs

1 a Graph D b Graph E c Graph B

d Graph F e Graph C f Graph A

2

$\frac{8}{x} = \frac{1}{2}x$, $16 = x^2$, $x = \pm 4$, $y = \pm 2$

A is (4, 2) and B is (−4, −2).

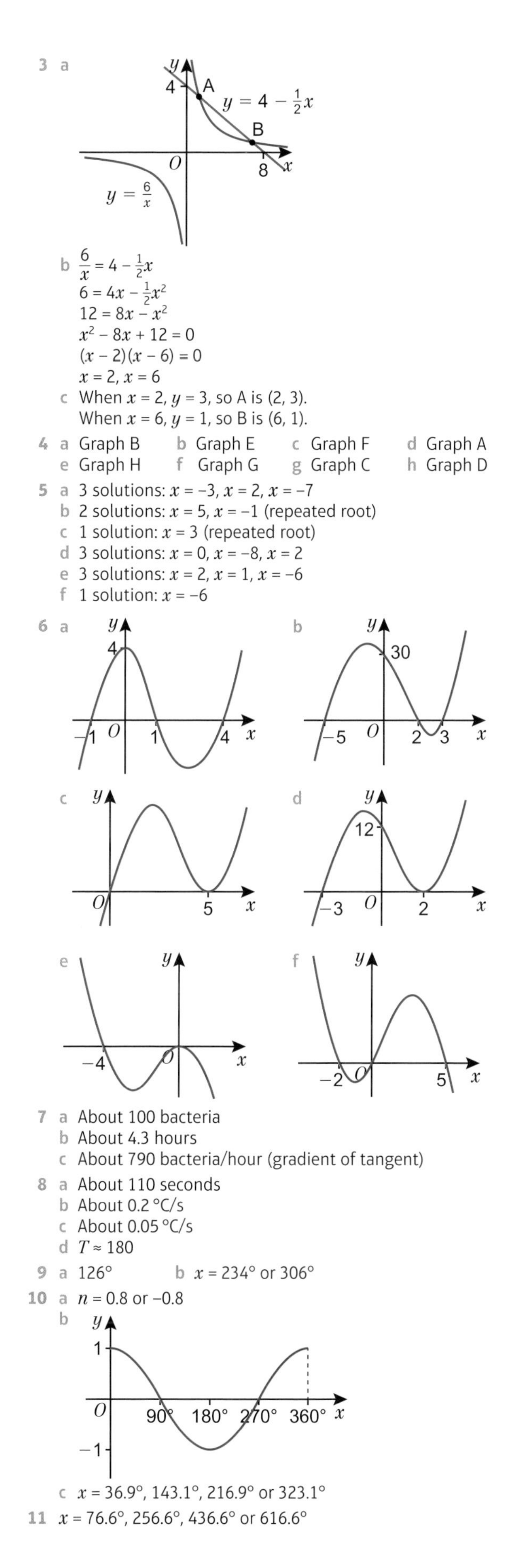

3 a 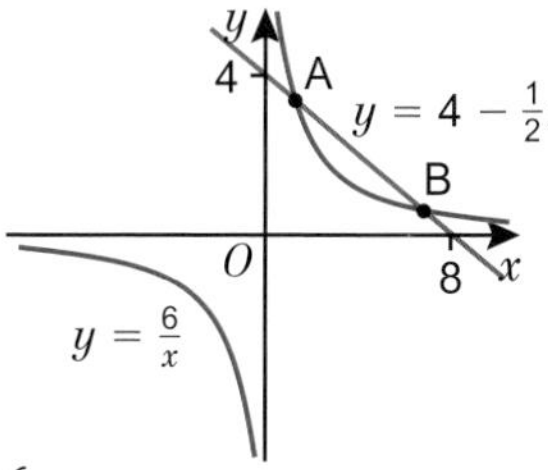

b $\frac{6}{x} = 4 - \frac{1}{2}x$

$6 = 4x - \frac{1}{2}x^2$

$12 = 8x - x^2$

$x^2 - 8x + 12 = 0$

$(x - 2)(x - 6) = 0$

$x = 2, x = 6$

c When $x = 2$, $y = 3$, so A is (2, 3).
When $x = 6$, $y = 1$, so B is (6, 1).

4 a Graph B b Graph E c Graph F d Graph A
e Graph H f Graph G g Graph C h Graph D

5 a 3 solutions: $x = -3, x = 2, x = -7$
b 2 solutions: $x = 5, x = -1$ (repeated root)
c 1 solution: $x = 3$ (repeated root)
d 3 solutions: $x = 0, x = -8, x = 2$
e 3 solutions: $x = 2, x = 1, x = -6$
f 1 solution: $x = -6$

6 a 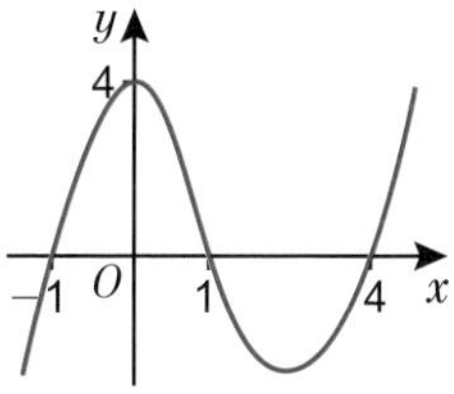

b 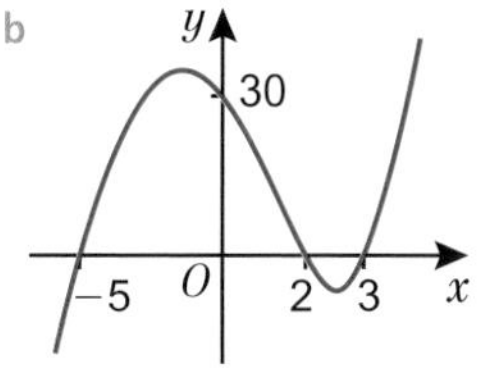

c 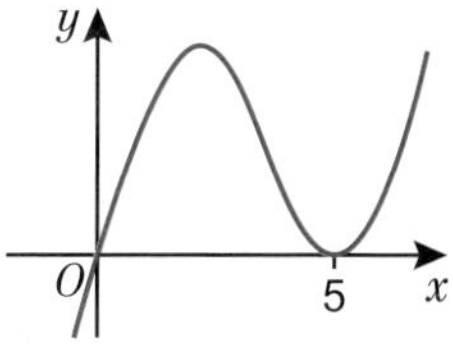

d 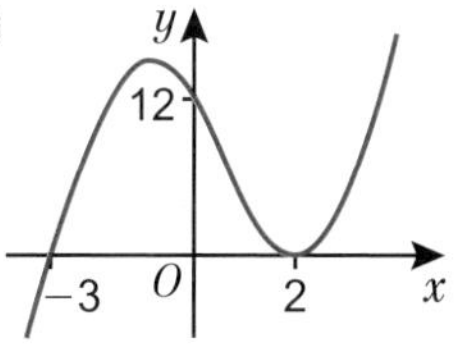

e 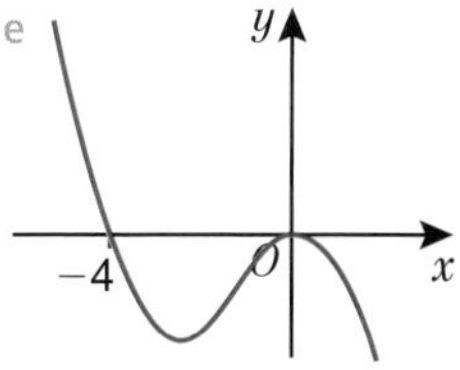

f 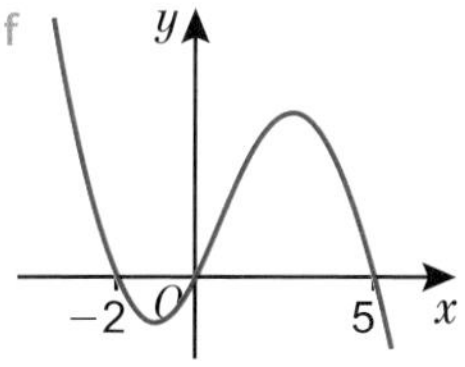

7 a About 100 bacteria
b About 4.3 hours
c About 790 bacteria/hour (gradient of tangent)

8 a About 110 seconds
b About 0.2 °C/s
c About 0.05 °C/s
d $T \approx 180$

9 a 126° b $x = 234°$ or 306°

10 a $n = 0.8$ or −0.8
b 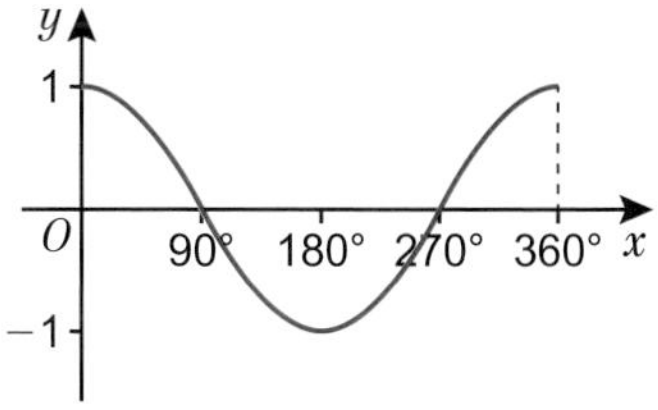

c $x = 36.9°, 143.1°, 216.9°$ or 323.1°

11 $x = 76.6°, 256.6°, 436.6°$ or 616.6°

2.7 Transformations of graphs

1 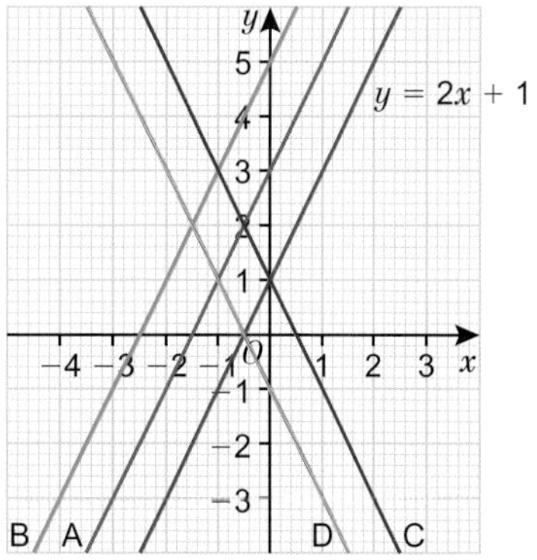

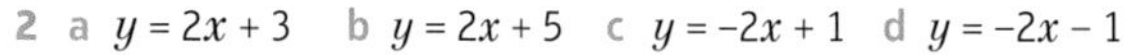

2 a $y = 2x + 3$ b $y = 2x + 5$ c $y = -2x + 1$ d $y = -2x - 1$

3 a i $2x + 3$ ii $2x + 5$ iii $-2x + 1$ iv $-2x - 1$
b The answers to Q2 and Q3 contain the same four expressions in x.

4 a, b 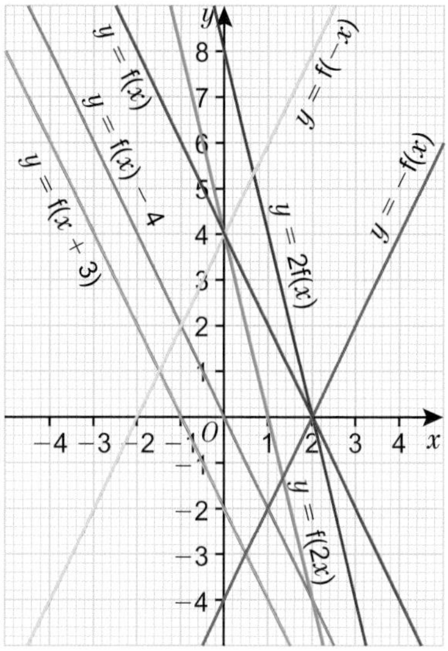

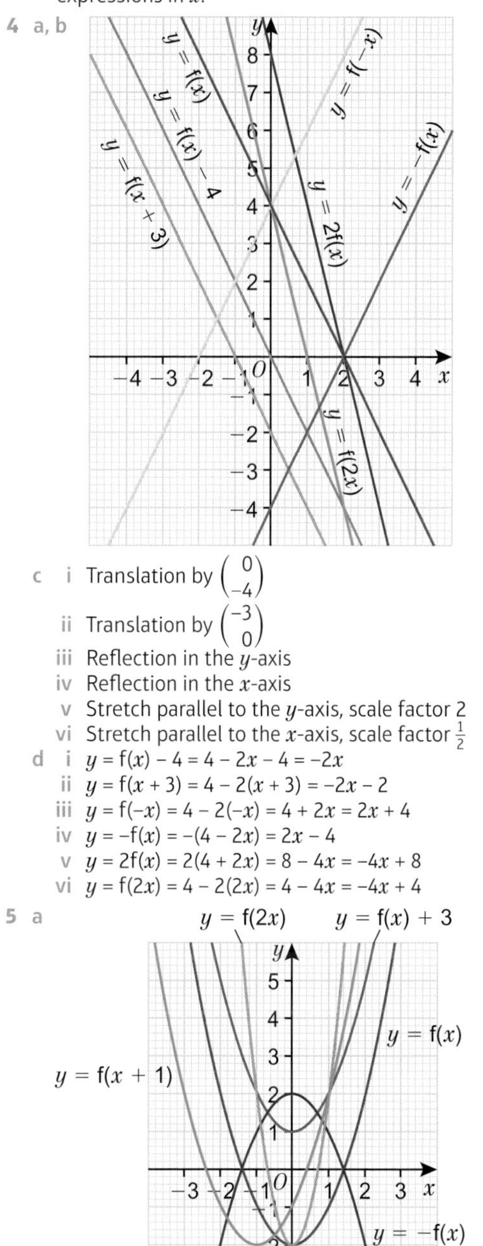

c i Translation by $\begin{pmatrix} 0 \\ -4 \end{pmatrix}$
ii Translation by $\begin{pmatrix} -3 \\ 0 \end{pmatrix}$
iii Reflection in the y-axis
iv Reflection in the x-axis
v Stretch parallel to the y-axis, scale factor 2
vi Stretch parallel to the x-axis, scale factor $\frac{1}{2}$

d i $y = f(x) - 4 = 4 - 2x - 4 = -2x$
ii $y = f(x + 3) = 4 - 2(x + 3) = -2x - 2$
iii $y = f(-x) = 4 - 2(-x) = 4 + 2x = 2x + 4$
iv $y = -f(x) = -(4 - 2x) = 2x - 4$
v $y = 2f(x) = 2(4 + 2x) = 8 - 4x = -4x + 8$
vi $y = f(2x) = 4 - 2(2x) = 4 - 4x = -4x + 4$

5 a 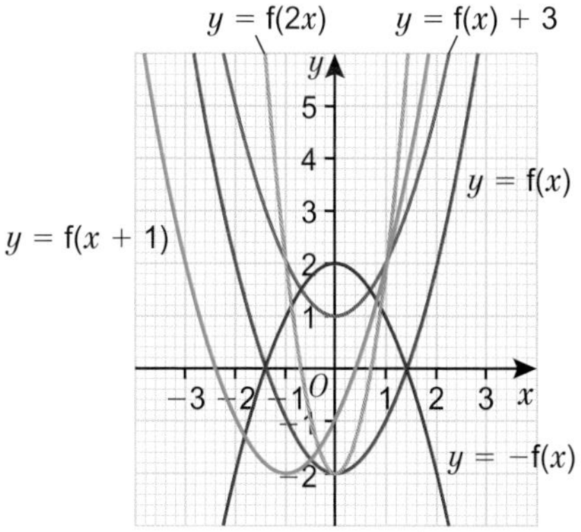

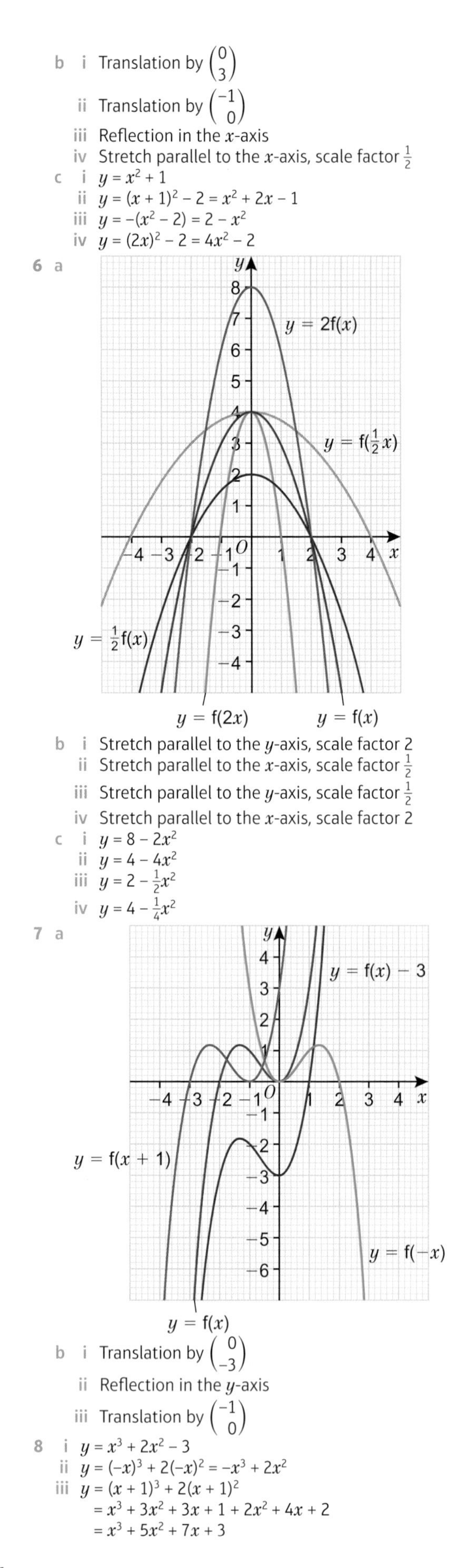

b i Translation by $\begin{pmatrix}0\\3\end{pmatrix}$

ii Translation by $\begin{pmatrix}-1\\0\end{pmatrix}$

iii Reflection in the x-axis

iv Stretch parallel to the x-axis, scale factor $\frac{1}{2}$

c i $y = x^2 + 1$

ii $y = (x + 1)^2 - 2 = x^2 + 2x - 1$

iii $y = -(x^2 - 2) = 2 - x^2$

iv $y = (2x)^2 - 2 = 4x^2 - 2$

6 a

b i Stretch parallel to the y-axis, scale factor 2

ii Stretch parallel to the x-axis, scale factor $\frac{1}{2}$

iii Stretch parallel to the y-axis, scale factor $\frac{1}{2}$

iv Stretch parallel to the x-axis, scale factor 2

c i $y = 8 - 2x^2$

ii $y = 4 - 4x^2$

iii $y = 2 - \frac{1}{2}x^2$

iv $y = 4 - \frac{1}{4}x^2$

7 a

b i Translation by $\begin{pmatrix}0\\-3\end{pmatrix}$

ii Reflection in the y-axis

iii Translation by $\begin{pmatrix}-1\\0\end{pmatrix}$

8 i $y = x^3 + 2x^2 - 3$

ii $y = (-x)^3 + 2(-x)^2 = -x^3 + 2x^2$

iii $y = (x + 1)^3 + 2(x + 1)^2$

$= x^3 + 3x^2 + 3x + 1 + 2x^2 + 4x + 2$

$= x^3 + 5x^2 + 7x + 3$

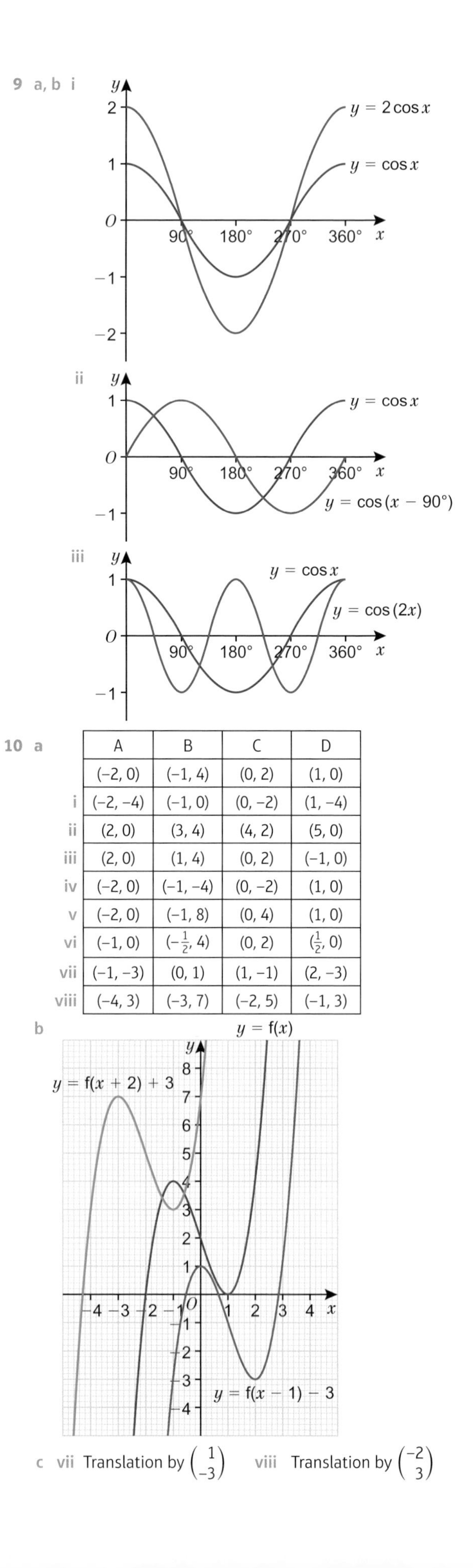

9 a, b i

ii

iii

10 a

	A	B	C	D
	(−2, 0)	(−1, 4)	(0, 2)	(1, 0)
i	(−2, −4)	(−1, 0)	(0, −2)	(1, −4)
ii	(2, 0)	(3, 4)	(4, 2)	(5, 0)
iii	(2, 0)	(1, 4)	(0, 2)	(−1, 0)
iv	(−2, 0)	(−1, −4)	(0, −2)	(1, 0)
v	(−2, 0)	(−1, 8)	(0, 4)	(1, 0)
vi	(−1, 0)	($-\frac{1}{2}$, 4)	(0, 2)	($\frac{1}{2}$, 0)
vii	(−1, −3)	(0, 1)	(1, −1)	(2, −3)
viii	(−4, 3)	(−3, 7)	(−2, 5)	(−1, 3)

b

c vii Translation by $\begin{pmatrix}1\\-3\end{pmatrix}$ viii Translation by $\begin{pmatrix}-2\\3\end{pmatrix}$

2.8 Simultaneous equations

1 a $(x+7)(x-7)$ b $2(x+10)(x-10)$ c $(x-9)(x+2)$ d $(3x-5)(x-4)$

2 a $y = 2x - \frac{3}{2}$ b $y = \frac{1}{3}x - 2$ c $y = \frac{3}{5}x - 4$ d $y = 2x - \frac{5}{2}$

3 a $x^2 - 8x + 16$ b $x^2 + 10x + 25$ c $49 - 14x + x^2$ d $15x^2 + 28x - 32$

4 a $x = -3, y = 9$ or $x = 4, y = 16$
b $x = -3, y = -15$ or $x = 8, y = 40$
c $x = \frac{3}{2}, y = -3$ or $x = 6, y = 6$

5 a $x = 4, y = 8$ or $x = -1, y = 3$
b $x = -\frac{5}{2}, y = -8$ or $x = 4, y = 5$
c $x = 6, y = 32$ or $x = -2, y = -8$

6 a $x = 0, y = -5$ or $x = 4, y = 3$
b $x = -\frac{5}{7}, y = -\frac{1}{7}$ or $x = -1, y = -1$
c $x = -\frac{7}{4}, y = -8$ or $x = 6, y = \frac{7}{3}$

7 $x^2 - 2x + 5 = 2x + 1$, so $x^2 - 4x + 4 = 0$, i.e. $(x-2)^2 = 0$
So there is only 1 distinct solution (when $x = 2$) and the graphs intersect at 1 point.
At (2, 5) the line is a tangent to the curve.

8 (0.90, 3.90) and (−3.90, −0.90)

9 $6 - 5x - x^2 = 10 - 2x$, so $x^2 + 3x + 4 = 0$
The discriminant, $b^2 - 4ac = 9 - 16 = -7$
$-7 < 0$ so there are 0 solutions.

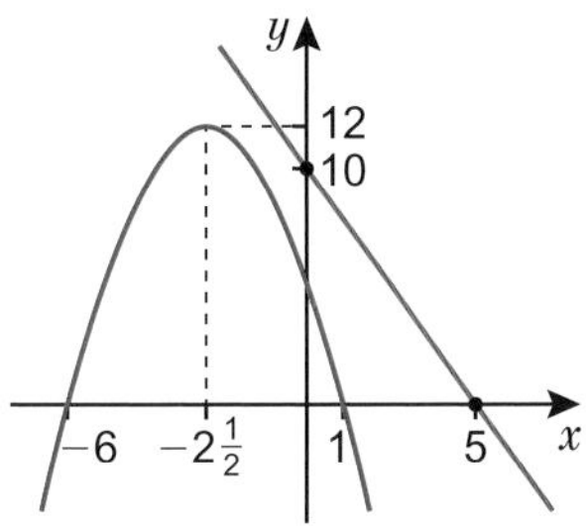

10 A is $(3 - \sqrt{6}, 6 + 2\sqrt{6})$ and B is $(3 + \sqrt{6}, 6 - 2\sqrt{6})$.

2.9 Equations, expressions and inequalities

1 a $x^2 - 5x - 36$ b $4x^2 - 28x + 49$ c $9x^2 + 12xy + 4y^2$

2 a $5(x-7)$ b $(2x+5y)(2x-5y)$ c $(x-5)(x-12)$ d $(2x-11)(x+4)$

3 a $\frac{11x}{12}$ b $\frac{9}{2a}$

4 a $x^2 + 8x + 15$ b $x^3 + 4x^2 - 17x - 60$ c $x^3 + 6x^2 - 13x - 42$

5 a $x^3 + 5x^2 - 12x - 36$ b $x^3 - 11x^2 + 20x + 32$ c $3x^3 - 25x^2 + 42x + 40$ d $6x^3 - 11x^2 - 26x + 40$

6 a $\frac{1}{5}$ b $\frac{x+6}{x-1}$ c $\frac{x}{x+3}$

7 a $x^2 + 6x$ b $4x$ c $\frac{3}{x+6}$ d $\frac{x-5}{x+10}$ e $\frac{x-8}{x+1}$ f $\frac{x+6}{x+9}$

8 a $\frac{2x-5}{3x-1}$ b $\frac{2x+1}{2x-5}$ c $\frac{3x-8}{x-4}$

9 a $\frac{2(w+3)}{3(w+2)}$ b $\frac{2(m-2)}{m+3}$ c $\frac{1}{(x+3)(x+1)}$ d $3t - 2$

10 a $\frac{5x-16}{6}$ b $\frac{2+3x}{2xy}$ c $\frac{4y-2x}{x^2y}$ d $\frac{3x^2-2y^2}{4xy}$

11 a $\frac{4x+14}{(x+2)(x+4)}$ b $\frac{11x+3}{(x+3)(x-2)}$ c $\frac{x-23}{(x+1)(x-5)}$ d $\frac{13-3x}{(x-3)(x-1)}$

12 a $\frac{11x+21}{(2x+3)(x-3)(x+3)}$ b $\frac{2x+22}{(x+7)(x-3)(x+4)}$ c $\frac{4x+7}{(3x-4)(x-2)(x+1)}$

13 a $x = -\frac{5}{3}$ or 2 b $x = -5$ or 2 c $x = -\frac{7}{3}$ or 1 d $x = -\frac{1}{2}$ or 5

14 a $x = \frac{6}{7}$ or −3 b $x = \frac{1}{6}$ or 2 c $x = \frac{4}{9}$ or 3 d $x = \frac{3}{2}$ or 5

15 a $x = 3.16$ or −0.16 b $x = 8.07$ or 0.93

16 a $\{x : -4 \leqslant x \leqslant 15\}$ b $\left\{x : x < \frac{1}{2}\right\} \cup \{x : x > 2\}$ c $\left\{x : -1 < x < \frac{5}{4}\right\}$ d $\left\{x : x \leqslant \frac{4}{3}\right\} \cup \{x : x \geqslant 5\}$

17 a $\left\{x : x < -\frac{2}{3}\right\} \cup \left\{x : x > \frac{3}{4}\right\}$ b $\left\{x : -\frac{3}{5} \leqslant x \leqslant \frac{1}{2}\right\}$ c $\left\{x : \frac{2}{3} < x < \frac{5}{3}\right\}$ d $\left\{x : x \leqslant -\frac{7}{4}\right\} \cup \left\{x : x \geqslant \frac{4}{5}\right\}$

18 a $x^2 - 3x - 5 = 0$, so $x^2 = 3x + 5$ and $x = \sqrt{3x + 5}$
b $x = 4.19258$

19 a $x = 4.82843$ b $x = 7.12310$

20 a $351 = 39 \times 9$
b
$$\begin{aligned}100x + 10y + z &= 99x + x + 9y + y + z \\ &= 99x + 9y + (x + y + z) \\ &= 99x + 9y + 9 \\ &= 9(11x + y + 1)\end{aligned}$$
Hence $100x + 10y + z$ is divisible by 9.

2.10 Sequences

1 a 17, 28, 45 b 11, 18, 29 c 19, 31, 50

2 a 256, 1024, 4096 b −162, −486, −1458 c 10, −5, $2\frac{1}{2}$

3 −2, −7, −8, −5, 2

4 a 9, 12, 21, 33 b 13, 17, 30, 47, 77 c −4, 1, −3, −2, −5, −7 d −8, 3, −5, −2, −7, −9

5 $n, -\frac{1}{2}n, \frac{1}{2}n, 0, \frac{1}{2}n, \frac{1}{2}n, n$

6 a $r = 1.5$, 8th term = 136.69 (2 d.p.)
b $r = -\frac{1}{3}$, 8th term $= -\frac{1}{9}$
c $r = \sqrt{2}$, 8th term $= 24\sqrt{2}$

7 About 2018 fish

8 a $\sqrt{3}$ b 2 c $162\sqrt{3}$

9 a Dividing any term by the preceding term gives the value of the common ratio, so
$\frac{K}{K+6} = r$ and $\frac{K-2}{K} = r$
Hence $\frac{K}{K+6} = \frac{K-2}{K}$
b $K = 3$
c $r = \frac{K-2}{K} = \frac{1}{3}$, so 8th term $= 9 \times \left(\frac{1}{3}\right)^7 = \frac{1}{243}$

10 a
Sequence: 6, 9, 14, 21, 30
1st differences: +3, +5, +7, +9
2nd differences: +2, +2, +2

b
Sequence: 2, 14, 34, 62, 98
1st differences: +12, +20, +28, +36
2nd differences: +8, +8, +8

c
Sequence: 8, 25, 54, 95, 148
1st differences: +17, +29, +41, +53
2nd differences: +12, +12, +12

11 a $2n^2 + 3$ b $3n^2 + 5$ c $\frac{1}{2}n^2 + 6$

12 a Substituting $n = 1$ in the expression for u_n gives $u_1 = a + b + c = 6$
b $4a + 2b + c = 7$ and $9a + 3b + c = 12$
c $3a + b = 1$ and $5a + b = 5$ d $a = 2$ and $b = -5$
e $c = 9$ f $2n^2 - 5n + 9$

13 $5n^2 - 11n + 8$

2.11 Mixed exercise

1 $y = -2x + 9$

2 a $12m^2 + 8mt - 15t^2$ b $3(5x + 2y)(5x - 2y)$

3 a

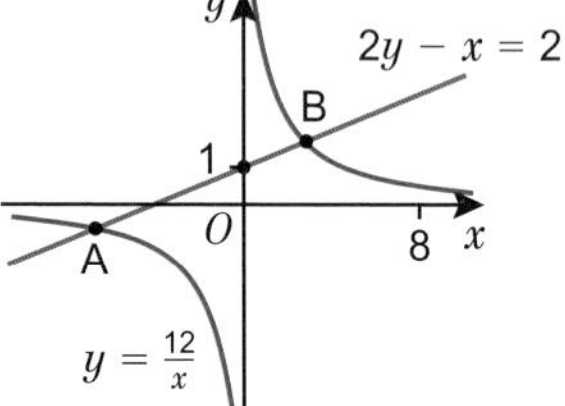

b A is (−6, −2) and B is (4, 3).

4 $y = -\frac{1}{2}x + \frac{3}{2}$

5 a

x	−2	−1	0	1	2	3	4
y	7	2	−1	−2	−1	2	7

b Graph of $y = x^2 - 2x - 1$ plotted using information in the completed table.
c $x = -1$ or 4

6 A and $y = x^2 + 4$, B and $y = x^3$, C and $y = 2^x$

7 a

x	−2	−1	0	1	2	3
y	18	5	−2	−3	2	13

b Graph of $y = 3x^2 - 4x - 2$ plotted using information in the completed table.
c $x = 1.7$ or -0.4
d Line $y = 5 - 2x$ added to graph; $x = -1.2$ or 1.9

8 a −7 b 8

9 a

x	−4	−3	−2	−1	0	1	2	3	4
y	0	7	12	15	16	15	12	7	0

b Graph of $y = 16 - x^2$ plotted using information in the completed table.
c 84 square units

10 a $\sqrt{52}$ b $y = \frac{3}{2}x + 13$

11 a 288° b $x = 108°$ or 252°

12 a $x = -\frac{1}{2}, y = 8$ or $x = 9, y = 27$
b $x = \frac{1}{4}, y = \frac{1}{2}$ or $x = 1, y = -1$

13 a $x^3 - 7x^2 - 14x + 120$ b $2x^3 + 7x^2 - 43x + 42$

14 a $x = 2, x = 6$ b $x = -7, x = 2$

15 $x = 4.16228$

16 a $3n^2 - 10n + 15$ b $4n^2 + 2n - 11$

17 a $a = 4, b = 5$ b (4, 5)

18 $x = 2.87, y = -0.87$ or $x = -0.87, y = 2.87$

19 $x = 6, y = 0$ or $x = -3.6, y = -4.8$

20 a $x = 4$ b $\frac{y - 15}{(y + 3)(y - 6)}$

21 a $2(x - 2)^2 + 5$ b Minimum point at (2, 5)

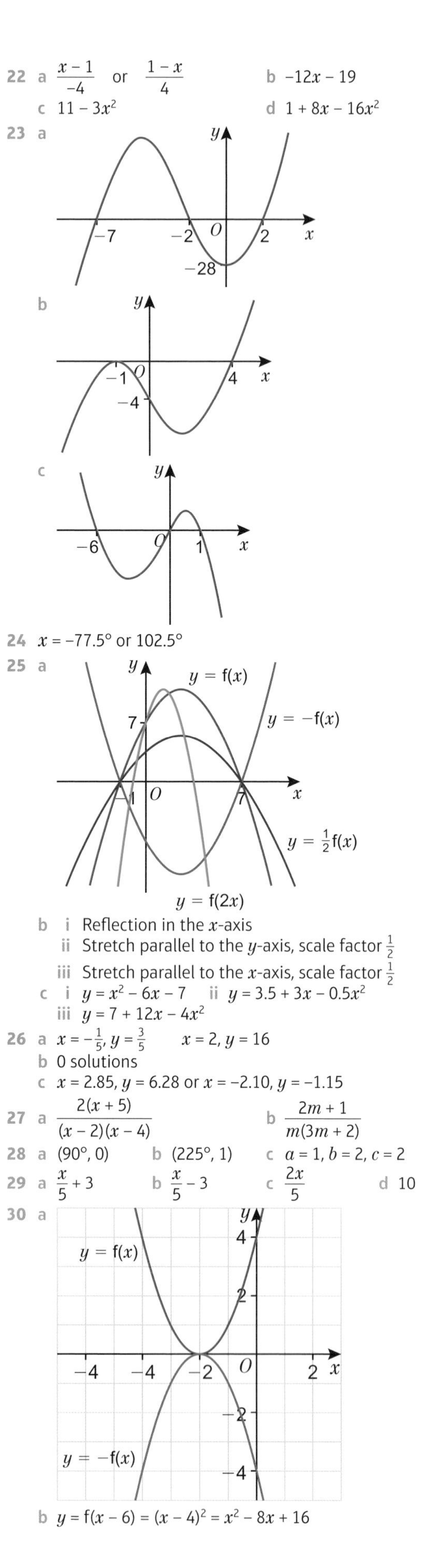

22 a $\frac{x - 1}{-4}$ or $\frac{1 - x}{4}$ b $-12x - 19$
c $11 - 3x^2$ d $1 + 8x - 16x^2$

23 a

b

c

24 $x = -77.5°$ or 102.5°

25 a

b i Reflection in the x-axis
ii Stretch parallel to the y-axis, scale factor $\frac{1}{2}$
iii Stretch parallel to the x-axis, scale factor $\frac{1}{2}$
c i $y = x^2 - 6x - 7$ ii $y = 3.5 + 3x - 0.5x^2$
iii $y = 7 + 12x - 4x^2$

26 a $x = -\frac{1}{5}, y = \frac{3}{5}$ $x = 2, y = 16$
b 0 solutions
c $x = 2.85, y = 6.28$ or $x = -2.10, y = -1.15$

27 a $\frac{2(x + 5)}{(x - 2)(x - 4)}$ b $\frac{2m + 1}{m(3m + 2)}$

28 a (90°, 0) b (225°, 1) c $a = 1, b = 2, c = 2$

29 a $\frac{x}{5} + 3$ b $\frac{x}{5} - 3$ c $\frac{2x}{5}$ d 10

30 a

b $y = f(x - 6) = (x - 4)^2 = x^2 - 8x + 16$

31 a Dividing any term by the preceding term gives the value of the common ratio, so:

$$\frac{k}{k+4} = r \quad \text{and} \quad \frac{2k-15}{k} = r$$

Hence $\frac{k}{k+4} = \frac{2k-15}{k}$

$k^2 = (2k-15)(k+4) = 2k^2 - 7k - 60$, so $k^2 - 7k - 60 = 0$

b $k^2 - 7k - 60 = (k+5)(k-12) = 0$, so $k = -5$ or 12

But we are told that k is positive, so $k = 12$

c $r = \frac{k}{k+4} = \frac{12}{16} = \frac{3}{4}$

32 $(2n+5)^2 - (2n+1)^2 = 4n^2 + 20n + 25 - (4n^2 + 4n + 1)$

$= 4n^2 + 20n + 25 - 4n^2 - 4n - 1$

$- 16n + 24$

$= 8(2n+3)$

Mean of the original integers $= \frac{2n+5+2n+1}{2}$

$= \frac{4n+6}{2}$

$= 2n + 3$

Hence the result.

33 $f(2n) = (2n)^2 - 3(2n) = 4n^2 - 6n$

$f(n+1) = (n+1)^2 - 3(n+1) = n^2 - n - 2$

$f(n+2) = (n+2)^2 - 3(n+2) = n^2 + n - 2$

LHS $= 2n^2 - 6n + 4$

$= 2(n^2 - 3n) + 4$

$= 2f(n) + 4$

$=$ RHS

3 RATIO AND PROPORTION

3.1 Compound measures

1 a 10.5 g/cm^3 **b** 46 cm^3

2 0.177 N/cm^2

3 39.05 kg

4 Largest distance = 137.825 m
Smallest distance = 128.625 m

5 Largest density = 3.55 g/cm^3
Smallest density = 3.43 g/cm^3

6 Largest mass = 16 362.5 tonnes
Smallest mass = 15 262.5 tonnes

7 Capacity of beaker: 750 ml to 850 ml
Volume of oil: 285 ml to 295 ml
Mass of iron bar: 3750 g to 3850 g
Density of iron bar: 7.85 g/cm^3 to 7.95 g/cm^3
Maximum volume of iron bar = 3850 ÷ 7.85 = 490.44… cm^3
Maximum volume in beaker (iron bar + oil)
= 490.44… + 295 = 785.44… cm^3
Minimum capacity of beaker = 750 cm^3, so the beaker might overflow.

3.2 Area and volume scale factors

1 a 69 cm **b** 28 cm^2

2 a 432 cm^3 **b** 180 cm^2

3 54 cm^2

4 a 1 : 6.25 **b** 8 cm^3

5 24.5 km^2

6 a 9.6 cm **b** 7 cm^2

7 10 buckets

8 1 : 1.4

9 a 6.5 cm^2 **b** 691.2 g

10 2763 m^2

3.3 Proportion

1 a The exchange rate for both is 1.39. So pounds and euros are in direct proportion.
b £1 = €1.39 **c** £190.50

2 80 N/m^2

3 a $y = 3.5x$ **b** $y = 38.5$ **c** $x = 12.8$
d Gradient = 3.5, y-intercept = 0

4 a $t = \frac{37}{3}\sqrt{m}$ $(= 12.33\sqrt{m})$ **b** $t = 148$ **c** $m = 2.25$

5 a $y = \frac{62.5}{x}$ **b** $y = 312.5$ **c** $x = \frac{5}{16}$ (= 0.3125)

6

p	2	**4**	6
w	7	$1\frac{3}{4}$	$\mathbf{\frac{7}{9}}$

7 a $h = \frac{15}{4}v^3$ $(= 3.75v^3)$ **b** $h = \frac{80}{9}$ (= 8.89) **c** $v = \frac{2}{3}$ (= 0.67)

8 a $y = \frac{1280}{x^2}$ **b** y is inversely proportional to x^2.
c $y = 80$ cm **d** $x = 8$ cm

9 a Graph B **b** Graph D **c** Graph A **d** Graph C

10 $w \propto \frac{1}{t^2}$ (rule B)

3.4 Exponential graphs

1 a $x = 7$ **b** $x = 0$ **c** $x = -2$ **d** $x = -4$

2 a

x	−3	−2	−1	0	1	2	3
y	8	**4**	**2**	**1**	0.5	**0.25**	0.125

b

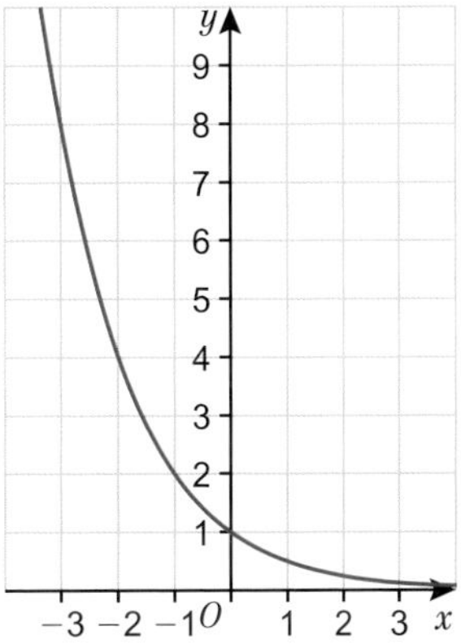

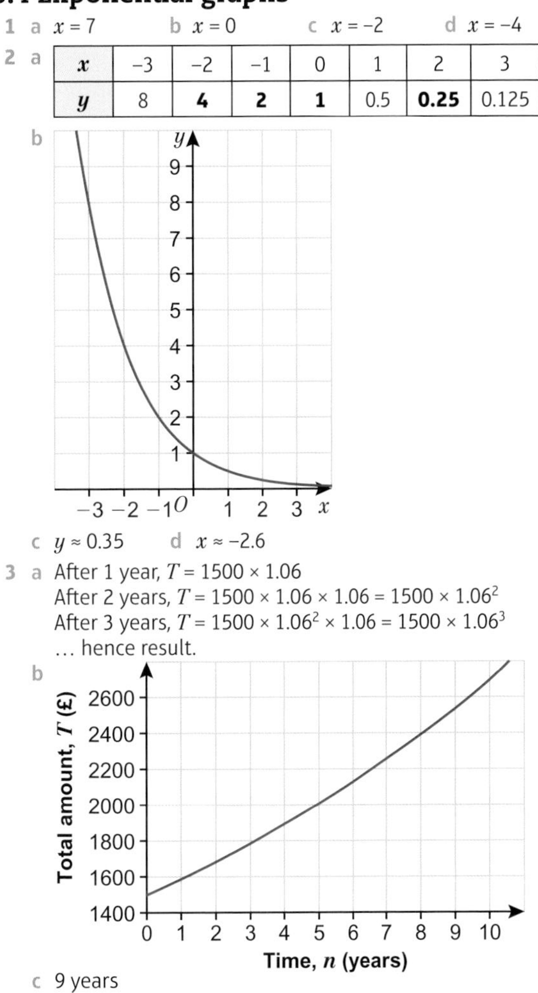

c $y \approx 0.35$ **d** $x \approx -2.6$

3 a After 1 year, $T = 1500 \times 1.06$
After 2 years, $T = 1500 \times 1.06 \times 1.06 = 1500 \times 1.06^2$
After 3 years, $T = 1500 \times 1.06^2 \times 1.06 = 1500 \times 1.06^3$
… hence result.

b

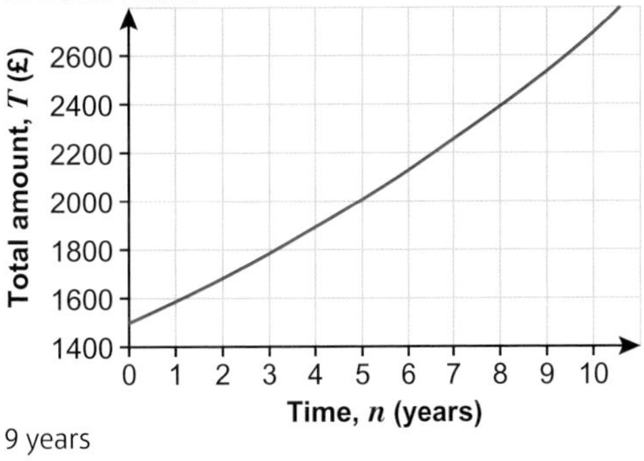

c 9 years

4 a

t	0	10	20	30	40	50	60	70	80
N	70	**35**	17.5	**8.75**	**4.4**	2.2	**1.1**	**0.55**	0.27

b

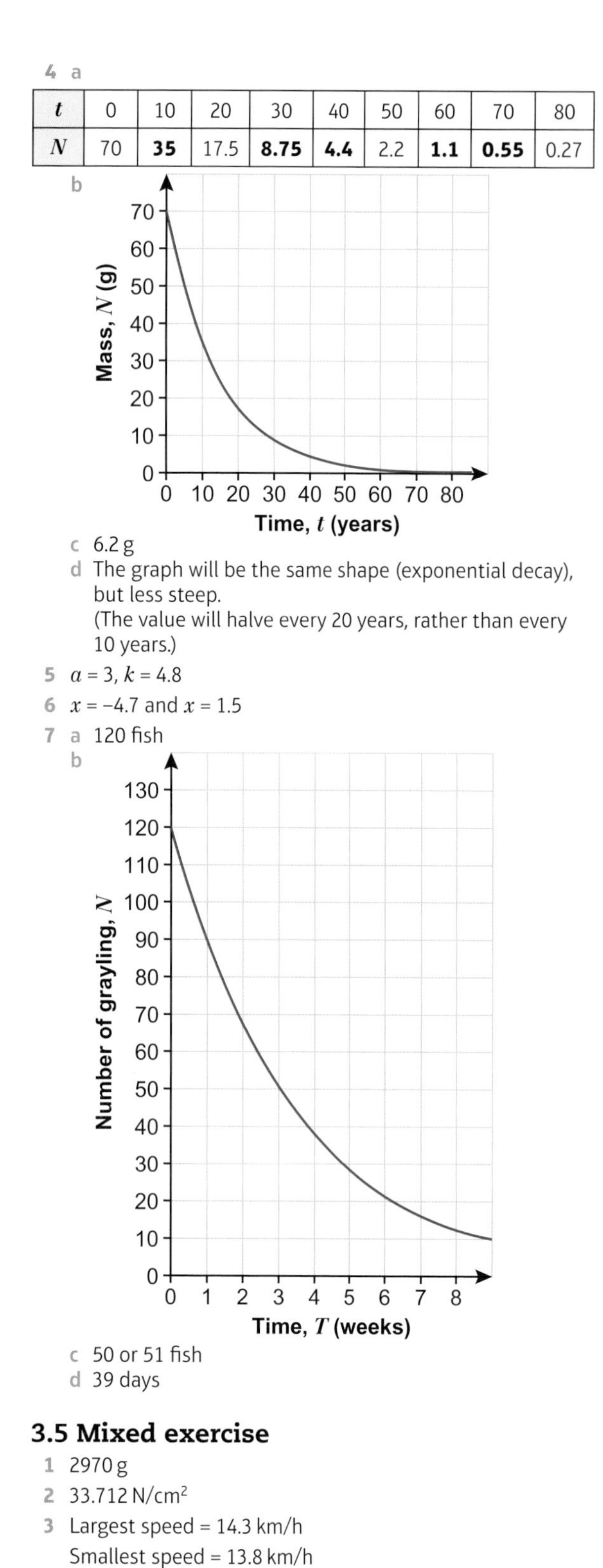

c 6.2 g

d The graph will be the same shape (exponential decay), but less steep.
(The value will halve every 20 years, rather than every 10 years.)

5 $a = 3$, $k = 4.8$

6 $x = -4.7$ and $x = 1.5$

7 a 120 fish

b

c 50 or 51 fish

d 39 days

3.5 Mixed exercise

1 2970 g

2 33.712 N/cm^2

3 Largest speed = 14.3 km/h
Smallest speed = 13.8 km/h

4 a $y = 4.5x^2$
b $y = 162$
c $x = 9$

5 44.8 litres

6 a

x	−3	−2	−1	0	1	2	3
y	**5.83**	3.24	**1.8**	**1**	0.56	**0.31**	0.17

b

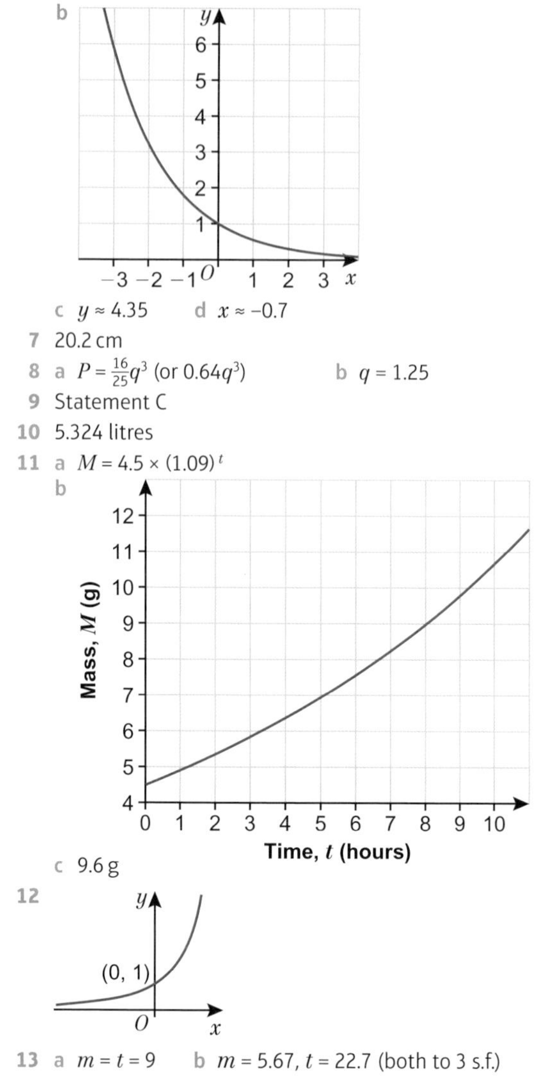

c $y \approx 4.35$ d $x \approx -0.7$

7 20.2 cm

8 a $P = \frac{16}{25}q^3$ (or $0.64q^3$) b $q = 1.25$

9 Statement C

10 5.324 litres

11 a $M = 4.5 \times (1.09)^t$

b

c 9.6 g

12

13 a $m = t = 9$ b $m = 5.67$, $t = 22.7$ (both to 3 s.f.)

4 PROBABILITY

4.1 Independent events and tree diagrams

1 a

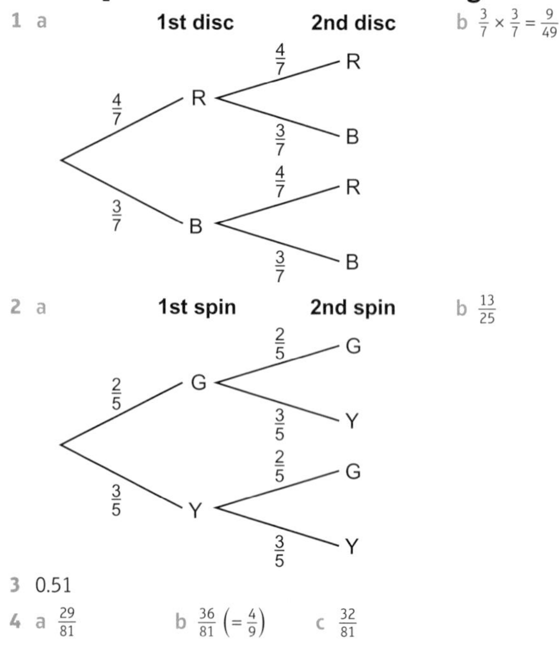

b $\frac{3}{7} \times \frac{3}{7} = \frac{9}{49}$

2 a

b $\frac{13}{25}$

3 0.51

4 a $\frac{29}{81}$ b $\frac{36}{81}\left(=\frac{4}{9}\right)$ c $\frac{32}{81}$

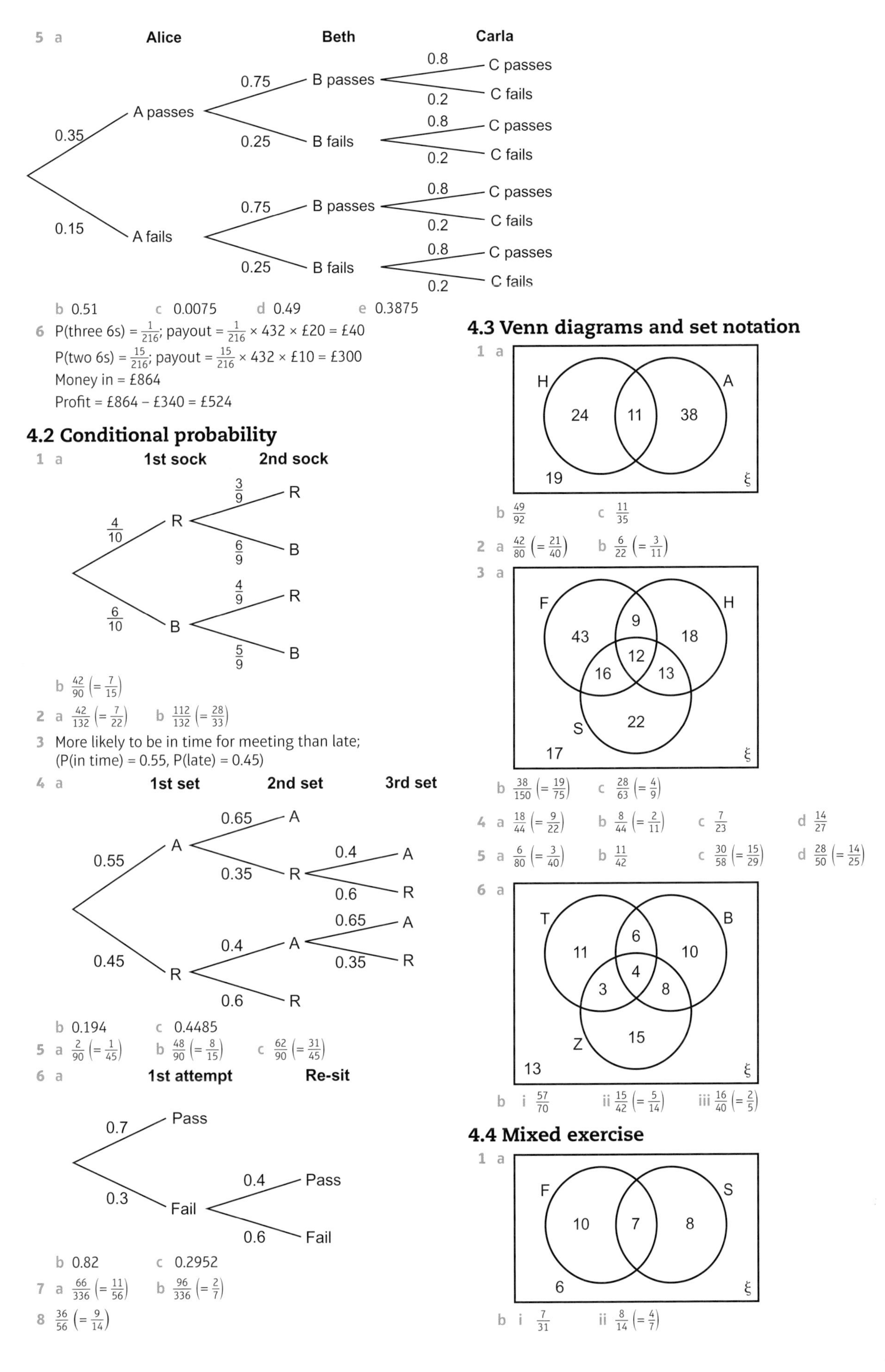

5 a

b 0.51   c 0.0075   d 0.49   e 0.3875

6 P(three 6s) = $\frac{1}{216}$; payout = $\frac{1}{216} \times 432 \times £20 = £40$
P(two 6s) = $\frac{15}{216}$; payout = $\frac{15}{216} \times 432 \times £10 = £300$
Money in = £864
Profit = £864 – £340 = £524

4.2 Conditional probability

1 a

b $\frac{42}{90}\left(=\frac{7}{15}\right)$

2 a $\frac{42}{132}\left(=\frac{7}{22}\right)$ b $\frac{112}{132}\left(=\frac{28}{33}\right)$

3 More likely to be in time for meeting than late; (P(in time) = 0.55, P(late) = 0.45)

4 a

b 0.194   c 0.4485

5 a $\frac{2}{90}\left(=\frac{1}{45}\right)$ b $\frac{48}{90}\left(=\frac{8}{15}\right)$ c $\frac{62}{90}\left(=\frac{31}{45}\right)$

6 a

b 0.82   c 0.2952

7 a $\frac{66}{336}\left(=\frac{11}{56}\right)$ b $\frac{96}{336}\left(=\frac{2}{7}\right)$

8 $\frac{36}{56}\left(=\frac{9}{14}\right)$

4.3 Venn diagrams and set notation

1 a

b $\frac{49}{92}$ c $\frac{11}{35}$

2 a $\frac{42}{80}\left(=\frac{21}{40}\right)$ b $\frac{6}{22}\left(=\frac{3}{11}\right)$

3 a

b $\frac{38}{150}\left(=\frac{19}{75}\right)$ c $\frac{28}{63}\left(=\frac{4}{9}\right)$

4 a $\frac{18}{44}\left(=\frac{9}{22}\right)$ b $\frac{8}{44}\left(=\frac{2}{11}\right)$ c $\frac{7}{23}$ d $\frac{14}{27}$

5 a $\frac{6}{80}\left(=\frac{3}{40}\right)$ b $\frac{11}{42}$ c $\frac{30}{58}\left(=\frac{15}{29}\right)$ d $\frac{28}{50}\left(=\frac{14}{25}\right)$

6 a

b i $\frac{57}{70}$ ii $\frac{15}{42}\left(=\frac{5}{14}\right)$ iii $\frac{16}{40}\left(=\frac{2}{5}\right)$

4.4 Mixed exercise

1 a

b i $\frac{7}{31}$ ii $\frac{8}{14}\left(=\frac{4}{7}\right)$

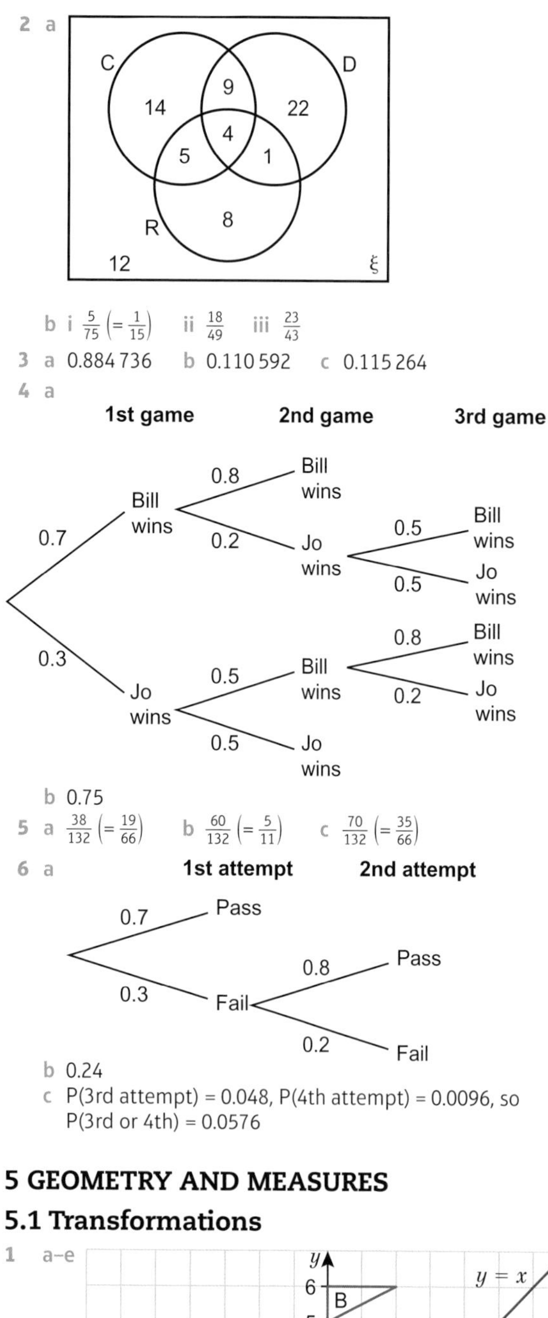

2 a

b i $\frac{5}{75}\left(=\frac{1}{15}\right)$ ii $\frac{18}{49}$ iii $\frac{23}{43}$

3 a 0.884 736 b 0.110 592 c 0.115 264

4 a

b 0.75

5 a $\frac{38}{132}\left(=\frac{19}{66}\right)$ b $\frac{60}{132}\left(=\frac{5}{11}\right)$ c $\frac{70}{132}\left(=\frac{35}{66}\right)$

6 a

b 0.24

c P(3rd attempt) = 0.048, P(4th attempt) = 0.0096, so P(3rd or 4th) = 0.0576

5 GEOMETRY AND MEASURES

5.1 Transformations

1 a–e

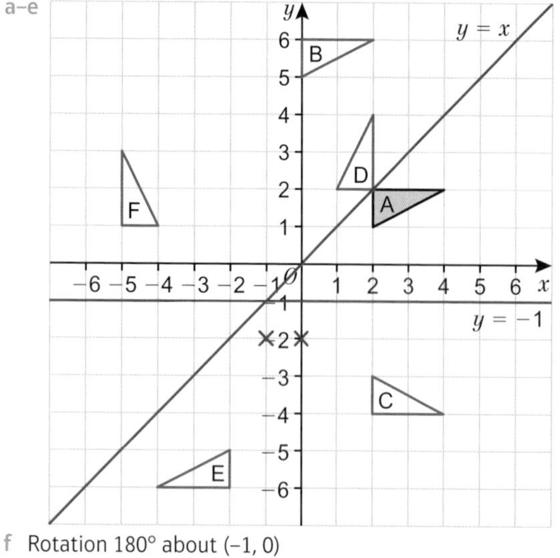

f Rotation 180° about (−1, 0)

g Rotation 90° clockwise about (−1, −1)

2 a–c

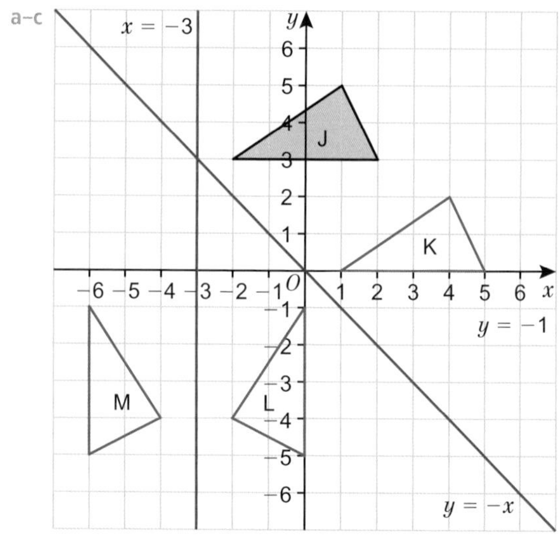

d Rotation 90° anticlockwise about (−6, 3)

3 a–e

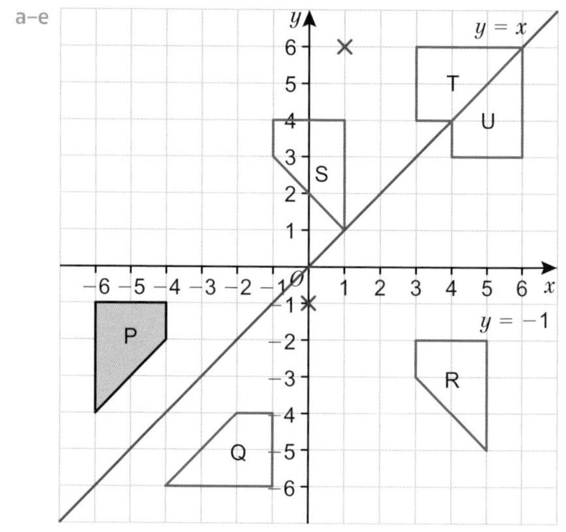

f Rotation 180° about (0, 1)

g Rotation 90° clockwise about (−5, 0)

h All the areas are the same. (Area is invariant under any combination of translation, reflection and rotation.)

4 a–d

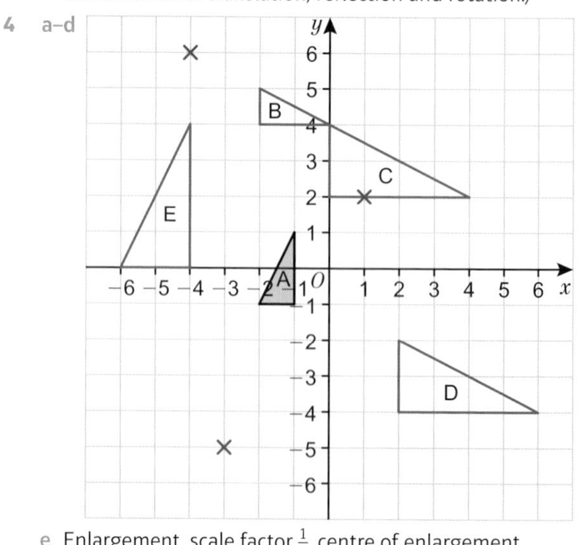

e Enlargement, scale factor 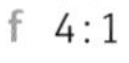 $\frac{1}{2}$, centre of enlargement (2, −2)

f 4 : 1

5 a, e

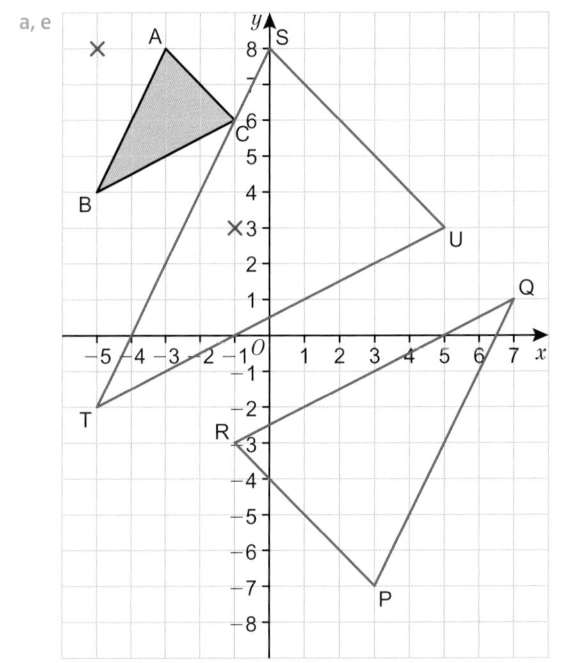

b Corresponding sides are parallel; each side on the image is twice as long as the corresponding side on the object.

c Enlargement, scale factor $-\frac{1}{2}$, centre of enlargement (–1, 3)

d 1:4

f 5:4 (or $2\frac{1}{2}$:2)

g 25:16

6 a–d

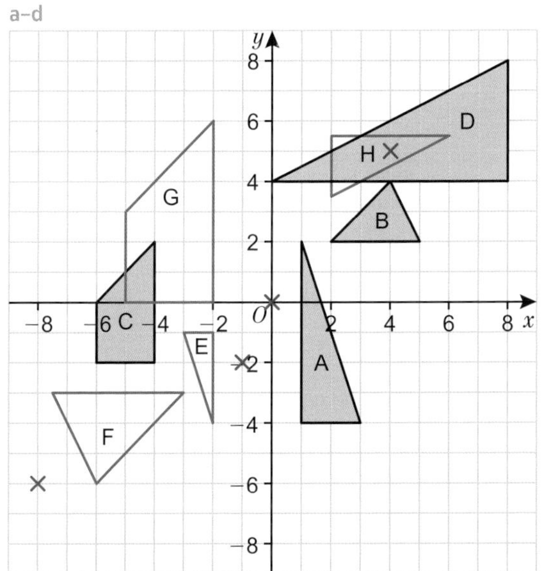

5.2 Area and volume

1 7.57 cm

2 Curved surface area = 1130 cm², volume = 679 cm³

3 a Arc length = 3.49 cm, area = 6.98 cm²
b Arc length = 38.7 cm, area = 145 cm²

4 a 72° b 15 cm c 45π cm²

5 a 107° b 12.9 cm c 29.5 cm²

6 a Volume = 209 cm³, curved surface area = 148 cm²
b Volume = 500 cm³, curved surface area = 264 cm²

7 a 351 cm² b 386 cm³

8 Surface area = 1520 cm², volume = 5580 cm³

9 545 cm³

10 Surface area = 206.25π cm², volume = 468.75π cm³

11 47.8 m³

12 Curved surface area = 140 cm², capacity = 185 cm³

5.3 Circle theorems

1 a Angle adjacent to 108° = 72° (angles on a straight line)
Two base angles of isosceles triangle are both 72°.
$x = 180 - 2 \times 72 = 36°$ (angle sum of triangle)
b Angle adjacent to 114° = 66° (angles on a straight line)
$x + 2x - 48 + 90 + 66 = 360°$ (angle sum of quadrilateral)
$x = 84°$
c $4x - 35 + x + 20 = 3x + 41$ (external angle property)
$x = 28°$

2 a Angle between tangent and radius = 90°
$x = 90 - 73 = 17°$ (angle sum of triangle)
b Angles between tangents and radii are both 90°.
$x + 90 + 90 + 46 = 360°$ (angle sum of quadrilateral)
$x = 134°$

3 a $x = 19°$ (base angle of isosceles triangle)
$y = 90 - 19 = 71°$ (angle between tangent and radius = 90°)
b $x = 63°$ (tangents of equal length, therefore base angle of isosceles triangle)
$y = 90 - 63 = 27°$ (angle between tangent and radius = 90°)

4 a $a = 119°$ (angle at the centre = 2 × angle at circumference)
b Reflex angle at centre = 220°
$b = 110°$ (angle at the centre = 2 × angle at circumference)
c $c = 52°$ (base angle of isosceles triangle)
$d = 104°$ (angle at the centre = 2 × angle at circumference, or exterior angle property)
d $e = 32°$ (base angle of isosceles triangle)
$f = 64°$ (angle at the centre = 2 × angle at circumference)
$g = 90 - e = 58°$ (angle in semicircle = 90°)
e $h = 36°$ (angles subtended by same arc)
$i = 44°$ (angle sum of triangle)
$j = 44°$ ($j = i$, angles subtended by same arc)
f Angles between tangents and radii are both 90°.
$k = 360 - 90 - 90 - 44 = 136°$ (angle sum of quadrilateral)
$l = 68°$ (angle at centre = 2 × angle at circumference)

5 Angle AOC = 2 × angle ABC (angle at centre = 2 × angle at circumference)
Therefore angle AOC = 60°
Triangle OAC is isosceles since OA = OC (radii)
Therefore angle OAC = angle OCA = $\frac{1}{2}(180 - 60) = 60°$
Therefore all angles in triangle OAC are 60°, so it is equilateral.

6 a $a = b = 40°$ (angles subtended by same arc)
Angle adjacent to 35° = 40° (alternate angles)
So $c = b + 75 = 115°$ (exterior angle property)
b Angle adjacent to d = 74° (opposite angles of a cyclic quadrilateral are supplementary)
$d = 106°$ (angles on a straight line)
$e = 102°$ (opposite angles of a cyclic quadrilateral are supplementary)
c $f = 65°$ (angles subtended by same arc)
$g = 102 - f = 37°$ (exterior angle property)
d $h = 90°$ (angle in a semicircle)
$i = 26°$ (angle sum of triangle)
$j = 116°$ (opposite angles of a cyclic quadrilateral are supplementary)

7 a Let angle BAO = x
Angle CDO = angle BAO = x (both subtended by arc BC)
Angle BOC = 2 × angle BAO = $2x$ (angle at centre = 2 × angle at circumference)
So angle BAO + angle CDO = $x + x = 2x$ = angle BOC
b Angle BAO = angle ABO (base angles of isosceles triangle OAB, equal radii)
Angle CDO = angle DCO (base angles of isosceles triangle OCD, equal radii)
Since angle BAO = angle CDO = x, all four angles = x

8 Angle ABC = 180° – x (angles on a straight line)
Also angle ABC = 180° – angle ADC (opposite angles of a cyclic quadrilateral are supplementary)
So angle ADC = x
Angle ADC + angle CDT = 180° (angles on a straight line)
So $x + y = 180°$

9 a Angle on straight line adjacent to 79° and 53° is 48°.
$a = 48°$ (alternate segment theorem)
b $b = 42°$ (angle sum of triangle)
$c = b = 42°$ (alternate segment theorem)
$d = 80°$ (alternate segment theorem, or angles on a straight line)
c $e = 65°$ (alternate segment theorem)
Angle between radius and chord = 90 – 65 = 25°
$f = 25°$ (angles subtended by same arc)
d Angles between tangents and chord opposite 62° angle are both 62° (alternate segment theorem)
$g = 180 - 62 - 62 = 56°$ (angle sum of triangle)
e $h = 83 - 42 = 41°$ (exterior angle property)
Angle adjacent to $i = 42°$ (alternate segment theorem)
$i = 55°$ (angles on a straight line)
f $j = 70°$ (alternate segment theorem)
$k = 58°$ (alternate segment theorem)
$l = 110°$ (opposite angles of a cyclic quadrilateral are supplementary)
$m = 180 - 58 - 58 = 64°$ (isosceles triangle formed by tangents of equal length)

10 Let angles BAC and DAC = x (angles subtended by same arc)
Angle BAD = 2 × angle BAC = $2x$ (CA bisects angle BAD)
Angle BOD = 2 × angle BAD = $4x$ (angle at centre = 2 × angle at circumference)
Angle DCT = angle DAC = x (alternate segment theorem)
Therefore angle DCT = $\frac{1}{4}$ angle BOD

11 Join AC.
Angle TAC = angle TCA = $90 - \frac{1}{2}x$ (triangle TAC is isosceles)
Angle TCA = angle ABC (alternate segment theorem)
So $90 - \frac{1}{2}x = x + 9$
$x = 54°$
(Joining A and C to O, the centre of the circle, and using angle at centre = 2 × angle at circumference, also leads to a solution.)

12 Angle OBA = 90° – $3x$ (angle between tangent and radius = 90°)
Angle OAB = 90° – $3x$ (base angle of isosceles triangle OAB, equal radii)
Angle OAC = x (base angle of isosceles triangle OAC, equal radii)
Therefore angle BAC = 90° – $3x$ + x = 90° – $2x$
Angle BOC = 2 × angle BAC = 180° – $4x$ (angle at centre = 2 × angle at circumference)
Angle TBO = angle TCO = 90° (angles between tangent and radii)
In quadrilateral TBOC, $y + 90 + 180 - 4x + 90 = 360°$
Therefore $y = 4x$
(There are at least four other ways to prove this result.)

5.4 Trigonometry

1 a 5.12 cm b 8.92 cm c 5.96 cm
2 a 39.8° b 72.3°
3 18π cm^2
4 a 3.38 cm^2 b 12.6 cm^2 c 7.71 cm^2
5 a 9.12 cm b 20.9 cm c 9.03 cm
6 108 cm^2
7 a 6.69 cm b 7.89 cm c 10.1 cm
8 a 42.3° b 76.1° c 22.2°
9 a 14.6 cm b 74.6° c 133 cm^2
10 a 8.16 cm b 10.6 cm c 8.34 cm
11 a 52.4° b 55.2° c 116°
12 a 30.4 km b 092°
13 a 10.3 km and 205° b 8.60 km
14 a 19.3 cm b 21.2°
15 a 18.7 m b 31.9° c 20.5°
16 a 40 cm b 18.1 cm c 42.2° d 24.2 cm e 48.6°
17 25.5°

5.5 Vectors

1 $x = \sqrt{40}$ cm, $y = \sqrt{56}$ cm
2 $\begin{pmatrix}10\\-7\end{pmatrix}$
3 a $\sqrt{106}$ b $\sqrt{53}$ c $\sqrt{153}$ d $\sqrt{205}$
4

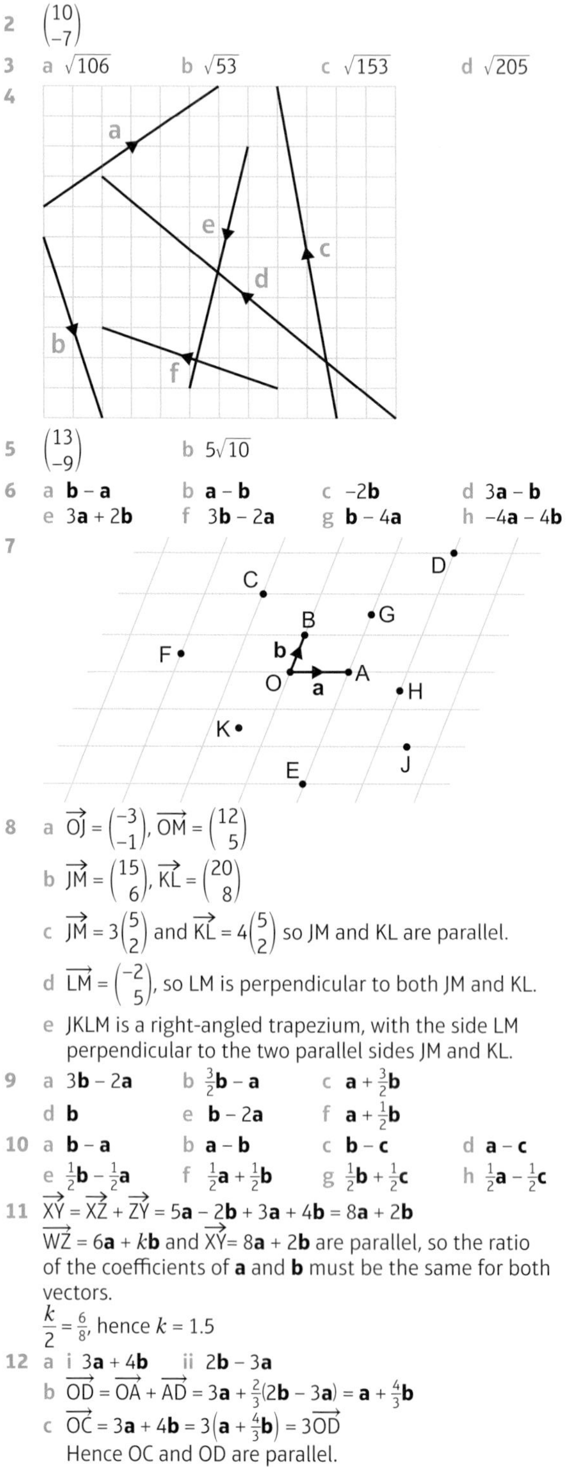

5 $\begin{pmatrix}13\\-9\end{pmatrix}$ b $5\sqrt{10}$
6 a **b** – **a** b **a** – **b** c –2**b** d 3**a** – **b**
e 3**a** + 2**b** f 3**b** – 2**a** g **b** – 4**a** h –4**a** – 4**b**
7

8 a $\overrightarrow{OJ} = \begin{pmatrix}-3\\-1\end{pmatrix}$, $\overrightarrow{OM} = \begin{pmatrix}12\\5\end{pmatrix}$
b $\overrightarrow{JM} = \begin{pmatrix}15\\6\end{pmatrix}$, $\overrightarrow{KL} = \begin{pmatrix}20\\8\end{pmatrix}$
c $\overrightarrow{JM} = 3\begin{pmatrix}5\\2\end{pmatrix}$ and $\overrightarrow{KL} = 4\begin{pmatrix}5\\2\end{pmatrix}$ so JM and KL are parallel.
d $\overrightarrow{LM} = \begin{pmatrix}-2\\5\end{pmatrix}$, so LM is perpendicular to both JM and KL.
e JKLM is a right-angled trapezium, with the side LM perpendicular to the two parallel sides JM and KL.

9 a 3**b** – 2**a** b $\frac{3}{2}$**b** – **a** c **a** + $\frac{3}{2}$**b**
d **b** e **b** – 2**a** f **a** + $\frac{1}{2}$**b**

10 a **b** – **a** b **a** – **b** c **b** – **c** d **a** – **c**
e $\frac{1}{2}$**b** – $\frac{1}{2}$**a** f $\frac{1}{2}$**a** + $\frac{1}{2}$**b** g $\frac{1}{2}$**b** + $\frac{1}{2}$**c** h $\frac{1}{2}$**a** – $\frac{1}{2}$**c**

11 $\overrightarrow{XY} = \overrightarrow{XZ} + \overrightarrow{ZY} = 5\mathbf{a} - 2\mathbf{b} + 3\mathbf{a} + 4\mathbf{b} = 8\mathbf{a} + 2\mathbf{b}$
$\overrightarrow{WZ} = 6\mathbf{a} + k\mathbf{b}$ and $\overrightarrow{XY} = 8\mathbf{a} + 2\mathbf{b}$ are parallel, so the ratio of the coefficients of **a** and **b** must be the same for both vectors.
$\frac{k}{2} = \frac{6}{8}$, hence $k = 1.5$

12 a i 3**a** + 4**b** ii 2**b** – 3**a**
b $\overrightarrow{OD} = \overrightarrow{OA} + \overrightarrow{AD} = 3\mathbf{a} + \frac{2}{3}(2\mathbf{b} - 3\mathbf{a}) = \mathbf{a} + \frac{4}{3}\mathbf{b}$
c $\overrightarrow{OC} = 3\mathbf{a} + 4\mathbf{b} = 3\left(\mathbf{a} + \frac{4}{3}\mathbf{b}\right) = 3\overrightarrow{OD}$
Hence OC and OD are parallel.
Since they also share the common point O, the points O, D and C are collinear, and ODC is a straight line.

8 a 250 runners

b 40 runners

c i

Time, t (minutes)	Frequency
$35 < t \leqslant 45$	18
$45 < t \leqslant 50$	54
$50 < t \leqslant 60$	110
$60 < t \leqslant 70$	40
$70 < t \leqslant 90$	28

ii 56.7 minutes

6.3 Mixed exercise

1 a Students' own answers, e.g. The range and the IQR are smaller at destination A so Paula is less likely to have extremes of temperature. The minimum temperature at A is higher than at B.

b Students' own answers, e.g. Destination B has a maximum of 36 °C and an UQ of 30 °C so if Paula likes really hot weather she might choose to go to B. The median is higher at B so, on average, B is warmer.

2 a

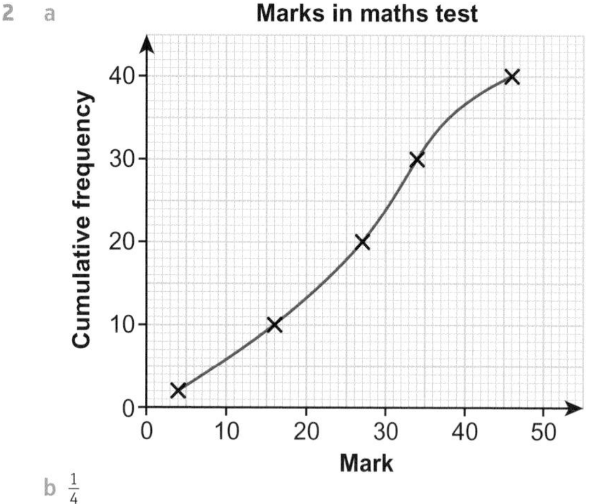

b $\frac{1}{4}$

3 a 52 babies b 3.2 kg c 3.7 − 2.5 = 1.2 kg

d The range and the IQR are greater for the Europe data, showing that the birthweights of these babies are more varied than in Asia. The median is greater for the Europe data, showing that, on average, the European babies are heavier at birth than the Asian babies.

4 a 68 seconds

b Yes, the captain is right, as 25% of the swimmers took less than 53 seconds and 53 < 60, or 28 swimmers took 60 seconds or less and 28 > 20.

c

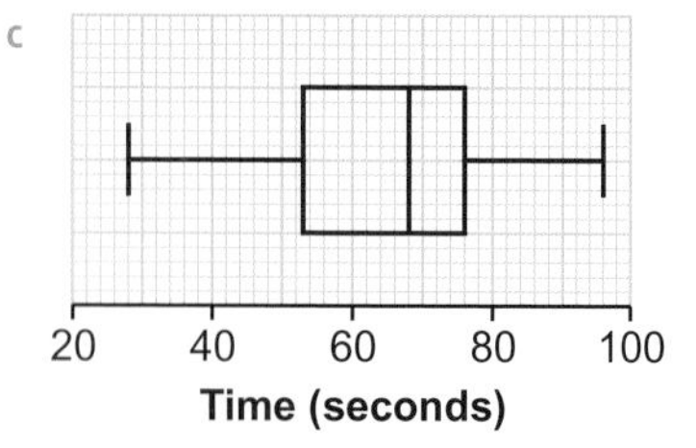

5

6 a

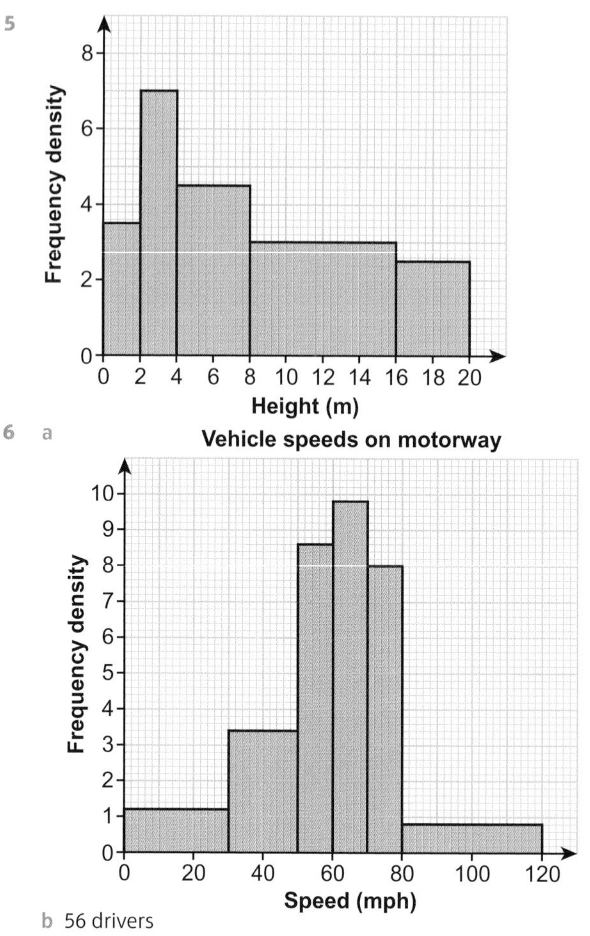

b 56 drivers

7 a Frequency for the $30 < t \leqslant 60$ class = 33

Bar for $150 < t \leqslant 210$ class drawn with a frequency density of 1.5

b 105 vehicles

8 About 86 or 87 farms

9 a Frequency for the $160 < h \leqslant 165$ class = 18

Frequency for the $165 < h \leqslant 170$ class = 25

Bar for $140 < h \leqslant 160$ class drawn with a frequency density of 1.1

Bar for $170 < h \leqslant 185$ class drawn with a frequency density of 1.8

Bar for $185 < h \leqslant 200$ class drawn with a frequency density of 0.8

b There are 104 adults in total, so the median height is the mean of the 52nd and 53rd heights.

The number of people up to 167 cm is 22 + 18 + 10 = 50 and the number of people up to 168 cm is 22 + 18 + 15 = 55.

So the median lies between 167 cm and 168 cm, and Clare is correct.

c 168.4 cm

6 STATISTICS

6.1 Comparison of distributions

1 a 15 b The median is still 15.

2 Any set of six numbers with a range of 8 and a total of 63, e.g. 7, 8, 10, 11, 12 and 15, or 6, 8, 10, 12, 13 and 14.

3 a 8 b 7 c 7 d 6.7

4 a i 26 minutes
ii 29 − 21 = 8 minutes
b **Time taken to complete puzzle**

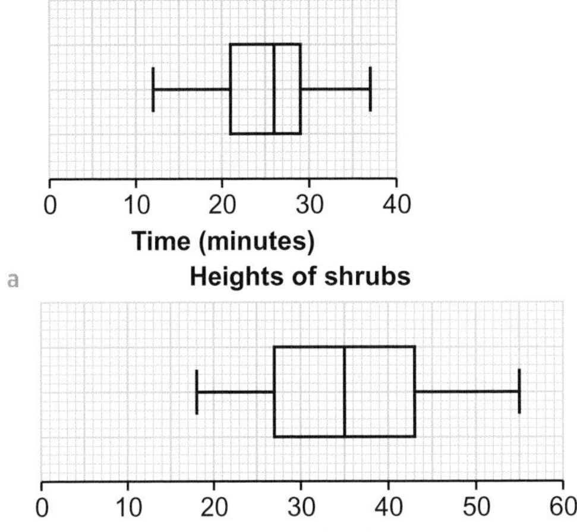

5 a **Heights of shrubs**

b Number of shrubs = 80 − 16 = 64,
percentage = $\frac{16}{80}$ × 100 = 20%
The cumulative frequency diagram, because it is impossible to use a box plot to estimate values in between the extreme values and quartiles that it shows, and 45 cm does not correspond to any of those.

6 a Students' own answers, e.g. The median for the boys is 5.8 hours and the median for the girls is 7.6 hours. So, on average, the girls spent longer on homework than the boys. The range for the boys (5.8) is less than for the girls (7.2) and the IQR for the boys (2.2) is less than for the girls (2.8), so the girls' times are more spread out than the boys' times.
b 25% of the boys and 50% of the girls did more than 7.6 hours.
Total = 0.25 × 32 + 0.5 × 26 = 8 + 13 = 21 students

7 a 80 tomatoes
b Minimum = 155 g, maximum = 205 g
c Median = 186 g, IQR = 191 − 176 = 15 g
d 100 tomatoes
e Minimum = 160 g, maximum = 210 g
f Median = 178 g, IQR = 183 − 174 = 9 g
g

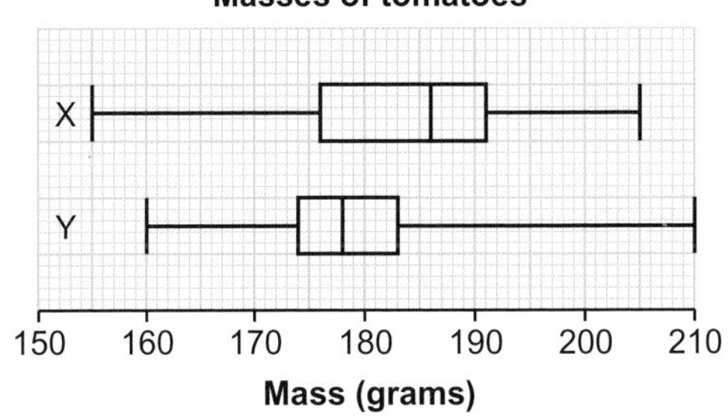

h The median for variety X (186) is greater than the median for variety Y (178) so, on average, tomatoes from variety X weigh more than tomatoes from variety Y. The range is the same for both varieties (50 g), but the IQR for variety Y (9) is significantly smaller than the IQR for variety X (15), so the masses of tomatoes from variety Y are more consistent.

8

	Town A	Town B
LQ	37	34
Median	45	40
UQ	53	56
IQR	16	22

The median for town B (40) is less than the median for town A (45) so, on average, the waiting times are shorter in town B. The IQR in town B (22) is greater than in town A (16), so the waiting times are more spread out in town B.

6.2 Histograms

1 a

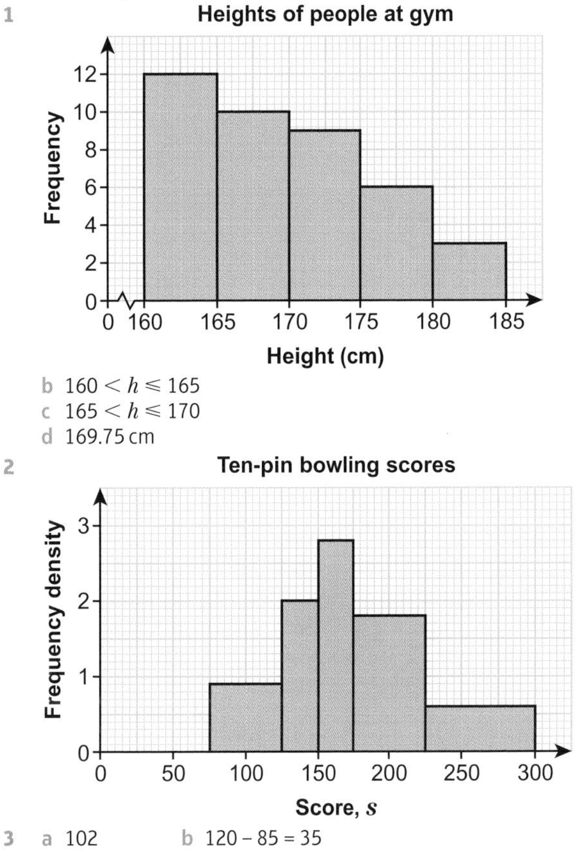

b $160 < h \leqslant 165$
c $165 < h \leqslant 170$
d 169.75 cm

2 **Ten-pin bowling scores**

3 a 102 b 120 − 85 = 35

4 a The number waiting for less than 20 minutes is 150. The total number of patients is 310, so Rachel is wrong.
b 38 patients

5 a Bar for $30 < t \leqslant 60$ class drawn with a frequency density of 1.8
Bar for $100 < t \leqslant 200$ class drawn with a frequency density of 2.5
Frequency for the $60 < t \leqslant 100$ class = 64
b 112 minutes

6 a 90 plants b 22 cm c 16 plants

7 a 23.8 g
b

Mass, m (grams)	Frequency
$16 < m \leqslant 20$	8
$20 < m \leqslant 22$	14
$22 < m \leqslant 24$	20
$24 < m \leqslant 25$	19
$25 < m \leqslant 26$	13
$26 < m \leqslant 29$	6

c 23.25 g

13 a $3\mathbf{b} - 3\mathbf{a}$
b $\overrightarrow{MP} = \overrightarrow{MA} + \overrightarrow{AP} = \overrightarrow{MA} + \frac{1}{3}\overrightarrow{AB} = \frac{3}{2}\mathbf{a} + \frac{1}{3}(3\mathbf{b} - 3\mathbf{a})$
$= \frac{3}{2}\mathbf{a} + \mathbf{b} - \mathbf{a}$
$= \frac{1}{2}\mathbf{a} + \mathbf{b}$
c $\overrightarrow{OQ} = \overrightarrow{OA} + \overrightarrow{AQ} = 3\mathbf{a} + \frac{2}{3}(3\mathbf{b} - 3\mathbf{a})$
$= \mathbf{a} + 2\mathbf{b}$
d OMPQ is a trapezium, with MP parallel to OQ and OQ = 2 × MP

14 a i $\overrightarrow{OQ} = \overrightarrow{OB} + \overrightarrow{BQ} = \mathbf{b} + 2\frac{1}{2}\mathbf{a}$
ii $\overrightarrow{AM} = \overrightarrow{AP} + \overrightarrow{PM} = \mathbf{b} - \frac{1}{2}\mathbf{a}$
iii $\overrightarrow{AN} = \frac{1}{3}\overrightarrow{AM} = \frac{1}{3}\left(\mathbf{b} - \frac{1}{2}\mathbf{a}\right)$
$= \frac{1}{3}\mathbf{b} - \frac{1}{6}\mathbf{a}$
iv $\overrightarrow{ON} = \overrightarrow{OA} + \overrightarrow{AN} = \mathbf{a} + \frac{1}{3}\mathbf{b} - \frac{1}{6}\mathbf{a}$
$= \frac{1}{3}\mathbf{b} + \frac{5}{6}\mathbf{a}$
b $\overrightarrow{ON} = \frac{1}{3}\mathbf{b} + \frac{5}{6}\mathbf{a}$
$= \frac{1}{6}(2\mathbf{b} + 5\mathbf{a})$
$\overrightarrow{OQ} = \mathbf{b} + 2\frac{1}{2}\mathbf{a}$
$= \frac{1}{2}(2\mathbf{b} + 5\mathbf{a})$
So $\overrightarrow{ON}$ and $\overrightarrow{OQ}$ are multiples of the same vector and hence are parallel. Since they also share the common point O, the points O, N and Q are collinear.

5.6 Mixed exercise

1 $x + 2x + 12 = 180°$ (opposite angles of a cyclic quadrilateral are supplementary)
Therefore $x = 56°$
$y = 112°$ (angle at the centre = 2 × angle at the circumference)

2 a–h

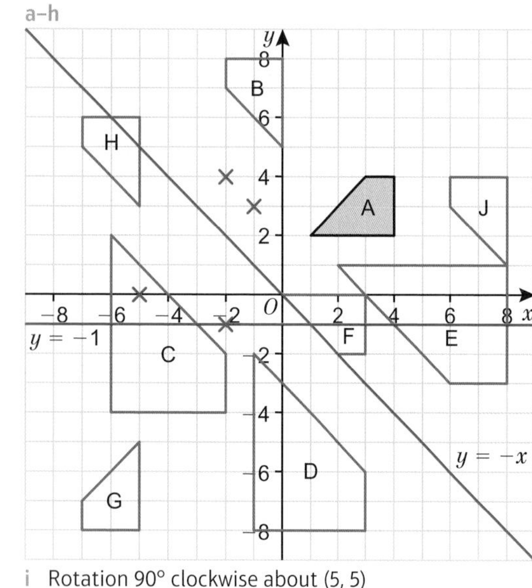

i Rotation 90° clockwise about (5, 5)

3 a 65 hours

4 Angle CBA = 73° (alternate segment theorem)
Angle CBO = $\frac{1}{2}(180 - 112) = 34°$ (base angle in isosceles triangle OBC)
Angle OBA = 73 − 34 = 39°

5 $V = 10x^2 + 24x - 18$

6

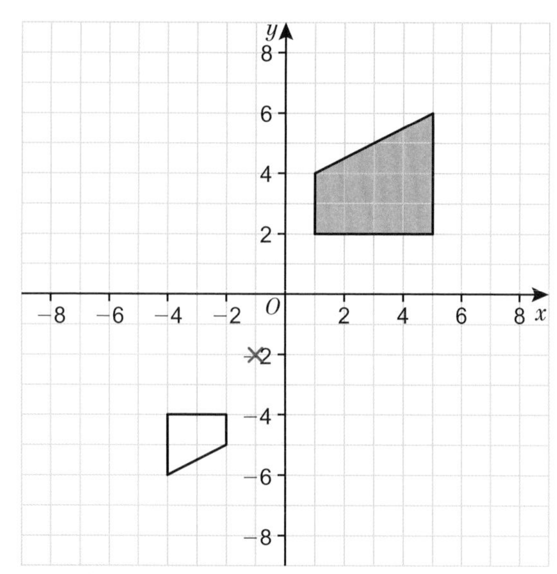

7 93.6 cm^2

8 DE = AE and AE = EB (tangents from an external point are equal in length), so DE = EB
AE = EC (given)
Therefore AE = DE = EB = EC, so DB = AC
The diagonals are equal and bisect each other, so quadrilateral ABCD is a rectangle.

9 $3x + 9x = 180°$ (opposite angles of a cyclic quadrilateral are supplementary)
$x = 15°$
Angle BED = 180 −15 −135 = 30° (angle sum of triangle BED)
Angle DCB = 30° (base angles of isosceles triangle BEC)
Angle BDC = 45° (angles on a straight line)
Therefore angle DBC = 105° (angle sum of triangle BDC)

10 $7\mathbf{a} - 5\mathbf{b}$

11 15 600 m^2 (3 s.f.)

12 a $\frac{3}{2}\mathbf{a}$
b MN is parallel to OA and is $\frac{3}{2}$ the length of OA.

13 $h = \frac{256}{81}x$

14 37.9 cm^2

15 18.6 cm

16 Area of each end face $= \frac{75}{360} \times \pi \times 144 = 30\pi$
Area of curved surface $= \frac{75}{360} \times 2 \times \pi \times 12 \times 20 = 100\pi$
Area of each rectangular face = 20 × 12 = 240
Total surface area $= 30\pi + 30\pi + 100\pi + 240 + 240$
$= 160\pi + 480$
$= 160(\pi + 3)$ cm^2

17 a i $-6\mathbf{a} + 4\mathbf{b}$
ii $-\frac{12}{5}\mathbf{a} + \frac{8}{5}\mathbf{b}$
iii $\frac{18}{5}\mathbf{a} + \frac{8}{5}\mathbf{b}$
iv $9\mathbf{a} + 4\mathbf{b}$
b $\overrightarrow{OR} = \frac{2}{5}(9\mathbf{a} + 4\mathbf{b}) = \frac{2}{5}\overrightarrow{OQ}$
So OR and OQ are parallel.
Since they also share the common point O, the points O, R and Q are collinear.

18 a 21.9 cm
b Angle ADB = 33.2°
c Angle DAE = 24.2°

19 a 204 cm^3
b 249 cm^2
c 70.5°